Einstein Studies

Editors: Don Howard John Stachel

Published under the sponsorship of the Center for Einstein Studies, Boston University

Volume 1:	Einstein and the History of General Relativity Don Howard and John Stachel, editors
Volume 2:	Conceptual Problems of Quantum Gravity Abhay Ashtekar and John Stachel, editors
Volume 3:	Studies in the History of General Relativity Jean Eisenstaedt and A.J. Kox, editors
Volume 4:	Recent Advances in General Relativity Allen I. Janis and John R. Porter, editors
Volume 5:	The Attraction of Gravitation: New Studies in the History of General Relativity John Earman, Michel Janssen and John D. Norton, editors
Volume 6:	Mach's Principle: From Newton's Bucket to Quantum Gravity Julian B. Barbour and Herbert Pfister, editors
Volume 7:	The Expanding Worlds of General Relativity Hubert Goenner, Jürgen Renn, Jim Ritter, and Tilman Sauer, editors
Volume 8:	Einstein: The Formative Years, 1879–1909 Don Howard and John Stachel, editors
Volume 9:	Einstein from 'B' to 'Z' John Stachel, editor

Don Howard John Stachel
Editors

Einstein
The Formative Years, 1879–1909

Springer Science+Business Media, LLC

Don Howard
Program in History and
Philosophy of Science
University of Notre Dame
Notre Dame, IN 46556
U.S.A.

John Stachel
Department of Physics
Center for Einstein Studies
Boston University
Boston, MA 02215
U.S.A.

Library of Congress Cataloging-in-Publication Data
Einstein: the formative years, 1879–1909 / Don Howard, John Stachel, editors.
p. cm. — (Einstein studies ; v. 8)
Includes bibliographical references and index.
ISBN 978-0-8176-3623-4
1. Einstein, Albert, 1879-1955 2. Physics–History–19th century. 3. Physicists–Biography. I. Series. II. Howard, Don, 1949- III. Stachel, John., 1928-

QC16.E5 E24 1998
530'.092—dc21 98-006996
CIP

Printed on acid-free paper.

Originally published by Birkhäuser Boston in 2000

Birkhäuser

SPIN 10647511
ISBN 978-0-8176-3623-4 ISBN 978-1-4612-0311-7 (eBook)
DOI 10.1007/978-1-4612-0311-7
Typeset by the editors in WordPerfect.

9 8 7 6 5 4 3 2 1

Contents

Preface vii
Acknowledgments ix
Contributors xi

Introduction to *Einstein: The Formative Years* 1
JOHN STACHEL

The Mysteries and Wonders of Natural Science: Aaron Bernstein's *Naturwissenschaftliche Volksbücher* and the Adolescent Einstein 23
FREDERICK GREGORY

The Young Einstein's Physics Education: H.F. Weber, Hermann von Helmholtz, and the Zurich Polytechnic Physics Institute 43
DAVID CAHAN

Kant's Impact on Einstein's Thought 83
MARA BELLER

Einstein's Controversy with Drude and the Origin of Statistical Mechanics: A New Glimpse from the "Love Letters" 107
JÜRGEN RENN

The Construction of the Special Theory: Some Queries and Considerations 159
ROBERT RYNASIEWICZ

Physical Approximations and Stochastic Processes in Einstein's 1905 Paper on Brownian Motion 203
SAHOTRA SARKAR

Einstein's Light-Quantum Hypothesis, or Why Didn't Einstein Propose a Quantum Gas a Decade-and-a-Half Earlier? 231
JOHN STACHEL

Index 253

Preface

This volume of *Einstein Studies* presents eight papers that offer new perspectives on many important themes concerning Albert Einstein's early life and work. The period covered—roughly the first thirty years of his life—comprises Einstein's years as a young student growing up in Munich, his subsequent studies at the Aargau Cantonal School in Aarau and the Swiss Federal Polytechnic Institute in Zurich, and his work as a clerk in the Swiss Federal Patent Office in Bern. Einstein's scientific activities during this period include his early work on the special theory of relativity, the quantum theory, and statistical physics, all of which are discussed in this volume. We have not included any discussion of the general theory of relativity, although his first steps toward that theory were taken in 1907. Rather than being the fruition of themes initiated during Einstein's formative years, this latter topic represents the commencement of new themes that were to dominate Einstein's middle and later years. These themes have been considered at length in other volumes of this series (see, especially, volumes 1, 3, 5, and 7).

Our aim in preparing this volume was to offer a selection of some of the best recent scholarly studies of Einstein's early years, most of them prepared especially for this occasion. All of these studies draw heavily upon the documentation regarding Einstein's early years that is now available in *The Collected Papers of Albert Einstein* (Princeton University Press), notably volumes 1, 2, and 5. The volume had its origin in a conference on "Einstein: The Early Years" that was held at Boston University's Osgood Hill Conference Center in October of 1990 under the auspices of The Boston University Center for Einstein Studies, a conference organized for the purpose of drawing attention to the new resources for Einstein scholarship being made available through the *Collected Papers*. The papers by Frederick Gregory, David Cahan, Mara Beller, and Robert Rynasiewicz are based on presentations at that meeting. To those authors and to the others who prepared their papers some time ago, we extend our sincere apologies for the delay in bringing this volume to press. One consequence of the delay is that, in a few places, the references are not as up to date as they might have been.

All of the papers are published here for the first time, with the exception of Jürgen Renn's contribution, which was first published in the *Archive for History of the Exact Sciences*. It is reprinted here with some changes made by the author.

Acknowledgments

The conference out of which the present volume grew was supported by The Center for Einstein Studies at Boston University. We thank Boston University and the staff of its former Osgood Hill Conference Center for helping to make the meeting such a pleasant occasion. To the participants in the meeting we extend our thanks for making it such an intellectually stimulating affair. Additional financial support from the Dibner Institute for the History of Science and the Dean S. Edmonds Foundation is gratefully acknowledged.

We also wish to thank Jed Buchwald, editor of the *Archive for History of Exact Sciences*, where Jürgen Renn's paper first appeared, for permission to reprint the paper here. The preparation of a volume such as this requires the work of many hands. Our sincere thanks go to our ever-patient and always-helpful editor at Birkhäuser, Ann Kostant. Further important help was provided at various stages by Yuri Balashov, Alisa Bokulich, and Patrick McDonald. To them all we extend our sincere thanks. We wish, lastly, to thank the contributors to this volume for their often-tried patience.

Contributors

Mara Beller, Department of Philosophy, The Hebrew University of Jerusalem, Mt. Scopus, 91 905 Jerusalem, Israel.

David Cahan, Department of History, 610 Oldfather Hall, University of Nebraska–Lincoln, Lincoln, NE 68588-0327.

Frederick Gregory, Program in History of Science, Department of History, University of Florida, Gainesville, FL 32611-7320.

Jürgen Renn, Max-Planck-Institut-für-Wissenschaftsgeschichte, Wilhelmstraße 44, 10117 Berlin, Germany.

Robert Rynasiewicz, Department of Philosophy, 348 Gilman Hall, Johns Hopkins University, Baltimore, MD 21218.

Sahotra Sarkar, Department of Philosophy, Waggener Hall 316, University of Texas at Austin, Austin, TX 78712-1180.

John Stachel, Department of Physics, Boston University, 590 Commonwealth Avenue, Boston, MA 02215.

Einstein

THE FORMATIVE YEARS, 1879–1909

Introduction to *Einstein: The Formative Years*

John Stachel

I.

In an attempt to characterize his scientific work as a whole, Einstein wrote: "The real goal of my research has always been the simplification and unification of the system of theoretical physics" (from the response to a questionnaire submitted to Einstein in 1932; Dukas and Hoffmann 1979, p. 11). Of course, such a sweeping retrospective assertion should be examined carefully before it is accepted; but, as we shall see, Einstein's earliest endeavors do provide considerable evidence of his effort to simplify and unify "the theoretical physicist's picture of the world" (Einstein 1918, p. 108), as he later called it,[1] an effort that came to dominate his outlook on physics. Fred Gregory (this volume, Chapter 2) gives us an insight into one of the popular-scientific sources of this effort, while Mara Beller (this volume, Chapter 4) gives us insights into one of its philosophical sources as well as into the ongoing regulative role it played in Einstein's work.

On occasion, Einstein's sometimes dazzling technical virtuosity was deployed to solve problems unconnected with his fundamental concerns. But by and large, his skills served his passion for a more profound understanding of the physical world. "From youth onwards my entire scientific effort was directed to deepening the foundations of physics" (from a recording made on 6 February 1924, in Herneck 1966). Sahotra Sarkar's description of Einstein's brilliantly simple technique for analyzing Brownian motion in order to provide direct evidence for atomism well illustrates this point (this volume, Chapter 7).

Einstein succinctly stated his criteria for judging a physical theory: "A theory is the more impressive the greater the simplicity of its premises, the more different kinds of things it relates, and the more extended its area of applicability" (Einstein 1979, p. 31). Such an increase in theoretical simplicity and comprehensiveness may be the result of any of several

Einstein Studies, vol. 8: Einstein: The Formative Years, pp. 1–22.

factors. First of all, it may result from a reduction in the number of fundamental concepts, or independent relations between these concepts.[2] The young Einstein was impressed, for example, by Hertz's attempt to eliminate the concept of force from the foundations of mechanics, and tried to emulate it in his own work.[3]

It also may result from developing a common theoretical explanation for a number of phenomena that previously appeared to be unrelated. As a student, Einstein was impressed by the quantitative relation between the coefficients of viscosity, heat conduction and diffusion established by the kinetic theory of gases.[4] As early as 1901, in commenting on his first paper, Einstein wrote: "It is a glorious feeling to recognize the unification of a complex of phenomena that appear to direct sense experience as completely separate things" (Einstein to Marcel Grossmann, 14 April 1901, Einstein 1987, Doc. 100).[5]

Finally, the increase may result from the development of a single, coherent theoretical structure, which embraces a number of less general theories that originally were independent of—or even in conflict with—each other. An outstanding example is Maxwell's theory (see Einstein 1979, pp. 30, 32), which subsumed the previously independent theories of electricity, magnetism, and optics; and which Einstein found "the most fascinating subject at the time I was a student" (Einstein 1979, p. 30).

Of course, the search for a unified theoretical basis for all of physics did not originate with Einstein. Three such foundational programs were current during his formative years, each with universal and exclusive pretensions: the traditional mechanical world-view based upon classical mechanics, the energeticist program based upon thermodynamics, and the electromagnetic world-view, based upon Maxwell's theory.[6]

From the time of Newton until the close of the nineteenth century, the prevalent view among physicists was that mechanical concepts ultimately would suffice to explain all physical phenomena, an outlook often referred to as the mechanical world-view.[7] While still quite young, Einstein became acquainted with this through the reading of several popular-scientific works.[8] His first scientific essay, written at the age of sixteen, takes for granted the possibility of a mechanical explanation of electromagnetic phenomena (see Einstein 1987, Doc. 5). He later recalled that, as an ETH student, he was profoundly impressed by the ability of mechanics to provide explanations "in areas that apparently had nothing to do with mechanics." As examples, he cites the mechanical theory of light, the kinetic theory of gases, and the deduction of the laws of thermodynamics from "the statistical theory of classical mechanics" (Einstein 1979, p. 18).

However, he adds that Mach's critical discussion of the foundations of mechanics, which he first read around 1897, helped protect him from dogmatic adherence to the mechanical world-view (see Section II).[9]

During the last decades of the nineteenth century, a few physicists and physical chemists claimed that it was possible to derive the laws of mechanics themselves from the principle of conservation of energy. By means of thermodynamical techniques, they hoped to base all of physics on the concept of energy, and hence referred to their program as energetics.[10] Skeptical of all forms of atomism, they were particularly critical of the kinetic theory of gases.[11] Einstein, who had adopted the atomistic outlook and accepted the kinetic theory while still a student (see Einstein 1987, the editorial notes "Einstein on Thermal, Electrical and Radiation Phenomena" and "Einstein on Molecular Forces"), does not appear to have been seriously influenced by this approach,[12] except negatively: The challenge to atomism spurred his search for more direct arguments for the existence of atoms and molecules.[13] The energeticist viewpoint, never widespread, did not long survive the theoretical attacks on its foundations around the turn of the century, and the new experimental evidence for atomism a few years later.[14]

By the time Einstein became acquainted with Maxwell's theory, physicists were gradually becoming accustomed to treating electric and magnetic fields as independent entities, to be employed—at least provisionally—on a par with mechanical ones.[15] When Maxwell formulated his theory, he still accepted the ultimate goal of a dynamical explanation of the electromagnetic field. The lack of any clear-cut success in the numerous attempts to provide such an explanation led by the turn of the century to skepticism about the entire enterprise. H.A. Lorentz's influential interpretation of Maxwell's theory greatly contributed to the acceptance of the field as an independent concept (see, for example, Lorentz 1895). Lorentz distinguished the physics of ponderable matter, composed of particles, some of which carry electric charges, from the physics of the ether, the exclusive seat of electromagnetic fields. The ether, present everywhere–even within matter–is totally immobile, does not interact mechanically with matter, but exerts electric and magnetic forces on charged particles. After the discovery of the electron, a charge carrier much less massive than any atom or ion, there were attempts to explain its inertial mass as only "apparent," that is, entirely due to the behavior of its electromagnetic field. The partial success of such attempts suggested the so-called electromagnetic world-view: the program of explaining all mechanical concepts and laws by means of Maxwell's theory.[16]

Yet however much he aspired to a unified approach, Lorentz was forced to employ a dualistic mixture of mechanical and electromagnetic concepts in order to formulate his theory of electrons. Einstein was presumably recalling his own acceptance at the turn of the century of such a dualistic outlook when he stated in 1914:

> About fifteen years ago, no one yet doubted that a correct account of the electrical, optical and thermal properties of matter was possible on the basis of Galilei-Newtonian mechanics applied to molecular motions and of Maxwell's theory of the electromagnetic field. (Einstein 1914, p. 740)

About the turn of the century, Einstein embarked on several lines of investigation that by 1905 led him to challenge the unrestricted validity of *both* Galilei-Newtonian mechanics and Maxwell's theory. But beyond his detailed criticisms of each of these theories, he was troubled by the dualism inherent in the contrast "between the theoretical concepts that physicists have formed regarding gases and other ponderable bodies and the Maxwell theory of electromagnetic processes in so-called empty space" (Einstein 1905a, p. 132). Einstein did not accept either the mechanical or the electromagnetic world-view as currently formulated; but his striving for simplicity and unification led him to anticipate the ultimate elimination of this dualism.[17] There is a discussion in Section III of his first, tentative proposals for its elimination; here we shall indicate some of his reasons for rejecting the unlimited validity of both Newtonian mechanics and Maxwell's electrodynamics.

Two problem complexes, each connected with the properties of electromagnetic radiation, led to these rejections. Einstein's early attempts to set up a consistent electrodynamics of moving bodies (see the editorial note in Einstein 1987, "Einstein on the Electrodynamics of Moving Bodies," pp. 223–225), were stymied by "the apparent incompatibility of the law of light propagation or rather of Lorentz's theory with the experimentally valid equivalence of all inertial systems" (from a recording made on 6 February 1924, in Herneck 1966, p. 134). In 1905 he succeeded in removing this apparent incompatibility by means of the theory of relativity. While Maxwell's equations are compatible with the principles of the theory, Einstein found that the Newtonian equations of motion for a particle had to be modified.

An even more "revolutionary" (Einstein to Conrad Habicht, 18 or 25 May 1905, Einstein 1993a, Doc. 27, pp. 31–32) challenge to the classical outlook emerged from Einstein's study of blackbody radiation (see the editorial note in Einstein 1987, "Einstein on Thermal, Electrical and Radiation Phenomena," pp. 235–237). Originally "stimulated by Wien's and Planck's investigations" at the turn of the century, by the middle of the decade Einstein had concluded "that mechanics and electrodynamics stand in insoluble contradiction to the facts of experience" (Herneck 1966, p. 134).[18]

Einstein was the first physicist to argue that the fundamental concepts of both mechanics and electrodynamics require drastic modification (see Einstein 1905a, Einstein 1906). But even though he regarded the dualistic conceptual basis he was forced to use as inadequate and provisional, and sought ways to modify the fundamental concepts to bring them into a harmonious unity, he was not disposed to prematurely jettison either the concepts of mechanics or of electrodynamics.[19] He was a master at finding his way among contradictory concepts of limited validity. He recognized that there are broad domains of phenomena, to the understanding of which classic mechanics and/or Maxwell's electrodynamics continue to provide reliable guidance.[20]

So he continued to investigate novel consequences of each theory within its domain of validity. In addition, he discovered ways to take advantage of contradictions between concepts by using one theory within its domain of validity to probe the limits of the other.

II.

Einstein's comments on foundational problems manifest an informed concern, which he (and others) later attributed in good part to his extensive readings in this area, as well as discussions with friends and fellow physicists. While a student, he read two of Ernst Mach's historical-critical studies, the *Mechanik* and *Wärmelehre* (see Einstein to Marić, 10 September 1899, Einstein 1987, Doc. 54, pp. 229–231, especially note 8), at the suggestion of his friend Michele Besso, with whom he discussed foundational questions.[21] In 1909, Einstein wrote to Mach:

> Naturally I know your major works quite well, among which I particularly admire the *Mechanics*. You have had such an influence on the epistemological outlook of the younger generation of physicists, that even your present-day opponents, such as Planck, without doubt would be regarded by one of those physicists, such as they were by-and-large a couple of decades ago, as "Machians." (Einstein to Ernst Mach, 9 August 1909, Einstein 1993a, Doc. 174, p. 204)

Commenting later on the prevailing mechanistic outlook in physics at this time, Einstein stated: "It was Ernst Mach who, in his *History of Mechanics*, upset this dogmatic faith; this book exercised a profound influence upon me in this regard while I was a student" (Einstein 1979, p. 18)"[22] The works of the "masters of theoretical physics" (Einstein 1955, p. 146) that he studied during his student years,[23] in particular those of Boltzmann, Helmholtz, Kirchhoff, and Hertz,[24] also contain sophisticated discussions of many foundational and methodological questions. His letters to her suggest that he discussed them with his fellow-student and fiancee, Mileva Marić. After graduation, Einstein appears to have discussed some of these questions with Alfred Kleiner, Professor of Physics at the University of Zurich, who became his *Doktorvater* and whose views may have influenced him.[25]

The wide-ranging "regular reading and discussion evenings" (Einstein to Michele Besso, 6 March 1952, Speziali 1972, p. 464) of Einstein and his friends Maurice Solovine and Conrad Habicht, who banded together in 1902 to form the playfully-named "Olympia Academy,"[26] undoubtedly played a significant role in the development of his ideas on foundational issues.[27] According to Solovine, "The end of the 19th and beginning of the 20th century was the heroic age of research on the foundations and principles of the sciences, and this is what was also our constant preoccupation" (Solovine 1956, p. viii). Solovine gave an extensive list of the readings that formed the basis of their discussions (Solovine 1956, pp. vii–viii).[28] He and Einstein started to read Karl Pearson's *Grammar of Science* (Pearson 1900) before Konrad Habicht joined them. After that, the three of them read and discussed: Mach's *Analyse der Empfindungen*[29] and *Mechanik*,[30] Mill's *Logic*,[31] Hume's *A Treatise of Human Nature*,[32] Spinoza's *Ethica* (Spinoza 1677),[33] unspecified memoirs and lectures by Helmholtz,[34] several chapters of Ampère's *l'Essai sur la philosophie des sciences* (Ampère 1834), Riemann's *Über die Hypothesen, welche der Geometrie zu Grunde liegen* (Riemann 1854), several chapters of Avenarius's *Kritik der reinen Erfahrung* (Avenarius 1888, 1890), Clifford's *On the Nature of Things-in-Themselves*,[35] Dedekind's *Was sind und was sollen die Zahlen?* (Dedekind 1893), and Poincaré's *La science et l'hypothèse* (Poincaré 1902).[36] Solovine describes Poincaré's book as one "that profoundly impressed us and kept us breathless for many weeks" (Solovine 1956, p. viii), and a number of references to Poincaré in Einstein's later writings suggest the influence of this careful reading (see, for example, Einstein 1917, § 31, p. 72; Einstein 1921, pp. 122– 123; Einstein to Michele Besso, 6 March 1952, Speziali 1972, p. 464).

Einstein's work on the theory of relativity constitutes a striking example of the fruitfulness of his epistemological concerns. The introduction to Einstein 1905b, his first paper on the subject, closes with a methodological comment that reflects these concerns:

> The theory to be developed here—like every other electrodynamics—is based upon rigid-body kinematics, since the assertions of any such theory concern relations between rigid bodies (coordinate systems), clocks and electromagnetic processes. Insufficient consideration of this circumstance is the root of the difficulties, with which the electrodynamics of moving bodies currently has to contend. (Einstein 1905b, p. 892)

By directing Einstein's attention to the problem of the formation of scientific concepts (including the concepts of space and time), the role of conventions in scientific thought, and the place of formal principles in the structure of scientific theories, to name only a few relevant topics, his readings and discussions helped prepare him for the task of revising the kinematical foundations of physics. Discussing "the axiom of the absolute character of time, or rather of simultaneity," Einstein later stated:

> To recognize clearly this axiom and its arbitrary character already implies the essentials of the solution of the problem. The type of critical reasoning required for the discovery of this central point was decisively furthered, in my case, especially by the reading of David Hume's and Ernst Mach's philosophical writings. (Einstein 1979, p. 50)[37]

Elsewhere, in citing readings of the Olympia Academy that were "rather influential in my development," Einstein listed Poincaré together with Hume and Mach (Einstein to Besso, 6 March 1952, Speziali 1972, p. 464.)

III.

Einstein's work during the period discussed in this volume may be classified under three broad headings:

(1) Efforts to extend and perfect the classical (Galilei-Newtonian) mechanical approach, including its statistical extensions (his work on molecular forces, the foundations of statistical physics, molecular dimensions, and Brownian motion). The papers of Sahotra Sarkar and Jürgen Renn deal in detail with this aspect of Einstein's work.
(2) Efforts to extend and perfect Maxwell's electrodynamics, as interpreted in Lorentz's electron theory, while modifying classical mechanics to make it cohere with electrodynamics (his work on the theory of relativity and the electrodynamics of moving media). Robert Rynasiewicz's paper deals in detail with this aspect of Einstein's work.

(3) Demonstrations of the limited validity of both classical mechanics and Maxwell's electromagnetic theory and efforts to comprehend phenomena that cannot be explained on the basis of these theories (his work on the quantum hypothesis). John Stachel's other paper discusses some aspects of this work.

Here, I shall outline Einstein's work in each of these categories. The works published through 1909 are all reprinted in Einstein 1989, and I include references to the very useful headnotes in that volume.

(1) *Efforts to Extend and Perfect the Classical-mechanical Approach*

By the turn of the century, Einstein was already occupied with the problems that were to take him beyond classical physics (see Section I). Yet his first five papers, published between 1901 and 1904, fall within the framework of Galilei-Newtonian mechanics and kinetic-molecular extensions of it. In the first two, Einstein attempted to explain several apparently disparate phenomena in liquids and solutions with the help of a simple hypothesis on the nature of intermolecular forces and their variation with the chemical composition of molecules (see Einstein 1989, editorial note, "Einstein on the Nature of Molecular Forces," pp. 3–8). Einstein hoped that his work might help decide a long-standing conjecture about a common basis for molecular and gravitational forces–an indication of his strong ambition, at the outset of his career, to contribute to the unification of physics.

Einstein made extensive use of thermodynamical arguments in the two papers, and such arguments played an important role in all his early work (see Klein 1967). The second paper raises a question about the relation between the thermodynamic and kinetic-molecular approaches to thermal phenomena, which is answered in the following paper, the first of three, published between 1902 and 1904, devoted to the atomistic foundations of thermodynamics (see Einstein 1989, editorial note, "Einstein on the Foundations of Statistical Physics," pp. 41–55). Einstein attempted to formulate the minimal assumptions about a mechanical system needed to derive the basic concepts and principles of thermodynamics. Presumably because he derived it from such general assumptions, he regarded the second law as a "necessary consequence of the mechanical world-view" (Einstein 1902, p. 432). He derived an equation for the mean square energy fluctuations of a system in thermal equilibrium that, in spite of its mechanical origins, involves only thermodynamical quantities. In his first mention in print of the topic, Einstein proceeded to apply this equation to black body radiation, the only system for which it was clear to him on what length scale such fluctuations should be significant. The application suggests that Einstein may have been attempting to treat radiation as a mechanical system.

In his doctoral dissertation, submitted in 1905, Einstein used classical-hydrodynamical methods to obtain an estimate of the size of large molecules in solution (see Einstein 1989, editorial note, "Einstein's Dissertation on the Determination of Molecular Dimensions," pp. 170–182). His papers on Brownian motion, the first of which appeared in the same year, also extend the scope of applicability of classical mechanical concepts (see Einstein 1989, editorial note, "Einstein on Brownian Motion," pp. 206–222). Einstein noted that, if the kinetic-molecular theory of heat is correct, the laws of thermodynamics cannot be universally valid, since fluctuations must give rise to microscopically visible violations of the second law. Indeed, such fluctuations provide an explanation of the well-known Brownian motion of microscopic particles suspended in a liquid. Einstein regarded his work as establishing the limits of validity, within which thermodynamics could be applied with complete confidence (see Einstein 1907c, p. 373).

(2) *Efforts to Extend and Perfect Maxwell's Electrodynamics and Modify Classical Mechanics to Cohere with It*

Well before 1905, Einstein apparently was aware of a number of experiments suggesting that the mechanical principle of relativity (the equivalence of all inertial frames for the description of mechanical phenomena) should be extended to optical and electromagnetic phenomena. However, such an extension was in conflict with what he regarded as the best current electrodynamical theory, Lorentz's electron theory, which grants a privileged status to one inertial frame: the ether rest frame. In 1905 Einstein succeeded in resolving this conflict through a critical analysis of the kinematical foundations of mechanics, which also underlie any theory of electrodynamics (see above). This analysis led to the recognition that the Lorentz transformations are kinematical in nature, and hence, any acceptable physical theory must be invariant under these transformations. Maxwell's equations, suitably reinterpreted after eliminating the concept of the ether, meet this requirement; but Newton's equations of motion need to be revised. Thus, Einstein's work on the theory of relativity provides an example of his use of one theory—Maxwell's electrodynamics—to find the limits of validity of another—Galileo-Newtonian mechanics, even though he was already aware of the limited validity of the former (see 3 below).

One of the major accomplishments of Einstein's approach, which his contemporaries found difficult to appreciate, is that relativistic kinematics is independent of the theories that impelled its formulation. Einstein had not only developed a coherent kinematical basis for both mechanics and electrodynamics, but (leaving aside the problem of gravitation) for any new physical theories that might be introduced. To use terms that he employed later, Einstein had created a theory of principle, rather than a constructive theory.[38] At the time he expressed the distinction in these words:

> One is in no way dealing here . . . with a "system," in which the individual laws would implicitly be contained and could be found merely by deduction therefrom, but only with a principle that (in a way similar to the second law of thermodynamics) permits the reduction of certain laws to others. (Einstein 1907b, p. 207)

The principles of such a theory, of which thermodynamics is his prime example, are universal assertions based upon a large amount of empirical data; they do not purport to constitute explanations of the phenomena on which they are based. In contrast, constructive theories, such as the kinetic theory of gases, do attempt to explain some class of phenomena, such as the gas laws on the basis of hypothetical entities employed to construct the explanation, such as atoms in motion. It is well known that elements of the distinction between principle and constructive theories occur in Poincaré's writings (see, for example, Poincaré 1904b).[39] Two other sources that may have inspired Einstein's emphasis on the role of principles in physics are Violle[40] and Kleiner.[41]

In spite of the merits of the theory of relativity, however, Einstein was of the opinion that

> a physical theory can only be satisfactory, if its structures are composed of *elementary* foundations. The theory of relativity is just as little ultimately satisfactory as, for example, classical thermodynamics was before Boltzmann had interpreted the entropy as probability. (Einstein to Arnold Sommerfeld, 14 January 190, Einstein 1993a, Doc. 73, pp. 86–89)[42]

Einstein's 1905 paper considered Maxwell's equations only with convection currents in empty space. The problem of formulating macroscopic, relativistically invariant equations for electromagnetic fields in moving media, including conduction currents, was first discussed by Minkowski in 1907 (see Minkowski 1908). In 1908, Einstein worked on this problem in collaboration with Jakob Laub (see Einstein 1989, editorial note, "Einstein and Laub on the Electrodynamcs of Moving Media," pp. 503–507).

(3) *Demonstrations of the Limited Validity of Both Classical Mechanics and Maxwell's Electromagnetic Theory and Attempts to Comprehend Phenomena that Cannot Be Explained by these Theories*

Einstein's efforts to perfect classical mechanics and Maxwell's electrodynamics, and to make both theories compatible, may still be regarded, in the broadest sense, as extensions of the classical approach to physics. However original his contributions in these areas, however revolutionary his conclusions about space and time appeared to his contemporaries, however fruitful his work proved to be for the exploration of new areas of physics, he was still engaged in drawing the ultimate consequences from conceptual structures that were well established by the end of the nineteenth century. What is unique about his stance during the first decade of this century is his unwavering conviction that the concepts of classical mechanics and those of Maxwell's electrodynamics, or even after his modifications of them, are incapable of explaining a growing list of phenomena involving the behavior and interactions of matter and radiation. Einstein constantly reminded his colleagues of the need to introduce radically new concepts to explain the structure of both matter and radiation. He introduced some of these new concepts himself, notably the quantum hypothesis, even though he was unable to integrate them into a coherent theory (see Einstein 1989, editorial note, "Einstein's Early Work on the Quantum Hypothesis," pp. 134–148).

His first paper on the quantum hypothesis provides a striking example of his style, mingling critique of the old concepts with the search for new ones. It opens by demonstrating that the equipartition theorem together with Maxwell's equations leads to a definite formula for the black body radiation spectrum, now known as the Rayleigh-Jeans distribution. This distribution, which at low frequencies matches the empirically-validated Planck distribution, cannot hold at all frequencies, since it implies a divergent total energy. (He soon gave a similar argument based on the equipartition theorem, demonstrating that classical mechanics cannot explain the thermal or optical properties of a solid, represented as a lattice of atomic or ionic oscillators; see Einstein 1907a.)

Einstein next investigated the high frequency (Wien) region, where the classically-derived distribution breaks down most dramatically. In this region, he showed, the entropy of a sample of monochromatic radiation with a fixed temperature depends on its volume in exactly the same way as does the entropy of a sample of a gas composed of statistically independent particles, no matter what their law of motion. In short, monochromatic radiation in the Wien limit behaves thermodynamically as if it were composed of mutually independent quanta of energy, each with an energy proportional to its frequency. Emboldened by this parallel, he took the final step, proposing the "very revolutionary" hypothesis that matter and radiation can only interact through the exchange of such energy quanta. He demonstrated that this hypothesis can be used to explain a number of apparently disparate phenomena involving the interaction of matter and radiation, notably the photoelectric effect.

In 1905, Einstein did not use Planck's distribution law to explore the nature of radiation. The following year, he showed that Planck's derivation of this law implicity depends on the assumption that charged oscillators can have only energies that are integral multiples of the quantum of energy, and hence only exchange energy with the radiation field in such quanta. In 1907, Einstein argued that uncharged oscillators should be similarly quantized, thereby explaining both the success of the DuLong-Petit law for most solids at ordinary temperatures, and the anomalously low values of the specific heats of certain substances. He related the temperature at which departures from the DuLong-Petit law become significant to the fundamental frequency of the atomic oscillators, and hence to the optical absorption spectrum of the solid (see Einstein 1907a).

In spite of his conviction of its fundamental inadequacy, Einstein continued to manipulate still-reliable aspects of classical mechanics with remarkable skill to explore further the structure of electromagnetic radiation. In 1909 he applied his theory of Brownian motion to a two-sided mirror immersed in thermal radiation. He showed that the mirror would be unable to carry out a Brownian motion indefinitely, if the fluctuations in the radiation pressure on its surfaces were solely due to the effects of random waves, as predicted by Maxwell's theory. Only the presence of an additional term, corresponding to pressure fluctuations due to the impact of random particles, guarantees the mirror's continued Brownian motion. Einstein showed that similar fluctuation terms in the energy are consequences of Planck's law. He regarded such fluctuation phenomena as his strongest argument for ascribing physical significance to the hypothetical light quanta.

Einstein was far from regarding his work on the quantum hypothesis as constituting a satisfactory theory of radiation or matter. As noted above, he emphasized that a physical theory is satisfactory only "if its structures are composed of *elementary* foundations," adding "that we are still far from having satisfactory elementary foundations for electrical and mechanical processes" (Einstein to Arnold Sommerfeld, 14 January 1908, Einstein 1993a, Doc. 73, pp. 86–89). Einstein felt that he had not reached a real understanding of quantum phenomena because (in contrast to his satisfactory interpretation of Boltzmann's constant as setting the scale of statistical fluctuations) he had been unable to interpret Planck's constant "in an intuitive way" (ibid.). The quantum of electric charge also remained "a stranger" to theory (see Einstein 1909a, p. 192). He was convinced that a satisfactory theory of matter and radiation should *construct* the quanta of electricity and of radiation, not simply *postulate* them.[43]

As a theory of principle (see above), the theory of relativity provides important guidelines in the search for such a theory. Einstein anticipated the ultimate construction of "a complete world-view that is in accord with the principle of relativity" (Einstein 1907c, p. 372). In the meantime, the theory offered clues to the constructions of such a world-view. One clue concerns the structure of electromagnetic radiation. Not only is the theory compatible with an emission theory of radiation, it requires that radiation transfer mass between an emitter and an absorber, reinforcing Einstein's hypothesis that radiation manifests a particulate structure under certain circumstances. He maintained that "the next phase in the development of theoretical physics will bring us a theory of light, which may be regarded as a sort of fusion of the undulatory and emission theories of light" (Einstein 1909b, pp. 482–483). Other principles that Einstein regarded as reliable guides in the search for an understanding of quantum phenomena are the conservation of energy and Boltzmann's principle.[44]

Einstein anticipated that: "The same theoretical modification that leads to the elementary unit [of charge] will also lead to the quantum structure of radiation as a consequence" (Einstein 1909a, pp. 192–193). In 1909 he made his first known attempt to find a field theory that would explain both the structure of matter (the electron) and of radiation (the light quantum). After investigating relativistically invariant, non-linear generalizations of Maxwell's equations, he wrote:

> I have not succeeded . . . in finding a system of equations, which I could see was suited to the construction of the elementary quantum of electricity and the light quantum. The manifold of possibilities does not seem to be so large, however, that one need draw back in fright from the task. (Einstein 1909b, p 193)[45]

This attempt may be regarded as the forerunner of his later search for a unified field theory of gravitation and electromagnetism, which would also explain quantum phenomena.

Einstein's attempt to incorporate gravitation into the theory of relativity led him to recognize a new formal principle, the principle of equivalence, which he interpreted as demonstrating the need to extend the relativity principle if gravitation is to be included in its scope (see Einstein 1989, editorial note, "Einstein on the Theory of Relativity," pp. 253–274).[46] He found that, when gravitational effects are taken into account, it is impossible to maintain the privileged role that the inertial frames of reference still hold in the original relativity theory (see Einstein 1907d, p. 454). He started to search for a group of transformations wider than the Lorentz group, under which the laws of physics would remain invariant when gravitation is included, a search that lasted until the end of 1915 and led to what he considered his greatest achievement, the general theory of relativity.

By 1909 then, Einstein at the age of 30 had already embarked on the search for a relativistic theory of gravitation, and begun the search for a theoretical foundation for quantum phenomena and a unified theory of matter and radiation, searches that were to occupy him for the rest of his life.

Notes

[1] For a discussion of the significance of the "Weltbild" concept for Einstein, see Holton 1981.

[2] "A theory obviously only gains scientific value through having simpler fundamental assumptions, i.e., of lesser multiplicity, than its consequences that are comparable with experience" (Einstein 1915, p. 254). Later, he spoke of "use of a minimum of primary concepts and relations."

[3] See Einstein 1979, pp. 28, 30. Einstein regarded a similar "elimination of the concept of force" as a major accomplishment of one of his first papers (Einstein to Michele Besso, 17 March 1903, Einstein 1993a, Doc. 7, p. 18).

[4] See, for example, Einstein 1979, p. 18. Einstein commented on the relation that the kinetic theory established between these quantities in 1901 (see Einstein to Mileva Marić, 15 April 1901, Einstein 1987, Doc. 101). In Einstein 1915, pp. 256–258, he discussed the three phenomena in some detail.

[5] This quotation shows a striking resemblance to the statement by Kant quoted by Beller (this volume, p. **xxx**).

[6] For a survey of these three programs, see Jungnickel and McCormmach 1986, Chap. 24, pp. 212–245.

[7] The term is used in Einstein 1902, p. 432. For his later account of the mechanical world-view, see Einstein 1979, pp. 16–30. For a discussion of the status of the mechanical outlook at the end of the nineteenth century, see Klein 1972.

[8] There is evidence that he read Alexander von Humboldt's *Kosmos,* Ludwig Büchner's *Kraft und Stoff,* and Aaron Bernstein's *Naturwissenschaftliche Volksbücher*. See Einstein 1987, editorial note on "Einstein's First Scientific Essay," pp. 5–6. For Aaron Bernstein, see Fred Gregory's article in this volume.

[9] Einstein presumably read the third edition of Mach's *Mechanik*, Mach 1897 (see Einstein 1987, p. 230, note 8). For his rereading of the *Mechanik*, see Section II, p. 7, and note 30 below.

[10] For a detailed survey of the energetics movement and its critics, see Deltete 1983.

[11] Not all critics of atomism were energeticists. Mach's critical stance, for example, was rooted in his general distrust of concepts not directly accessible to experience.

[12] In 1913, he wrote in praise of Planck's 1896 essay against the energeticists (Planck 1896), "in which it is shown that energetics is worthless as a heuristic method, indeed, that it even operates with untenable concepts" (Einstein 1913, p. 1077).

[13] He does not mention energetics in his reminiscences, although he does refer to the existence of skepticism about the reality of atoms (see Einstein 1979, pp. 44, 46), and to his goal, "to find facts that would guarantee as much as possible the existence of atoms of definite finite size" (Einstein 1979, p. 44).

[14] For a discussion of the theoretical attacks, see Deltete 1983. For a discussion of the experimental studies, see Nye 1972.

[15] For Einstein's later account of the challenge to the mechanical world-view presented by Maxwell's theory, see Einstein 1979, pp. 30–34. The German quotation in the text is from p. 30, the translation from p. 31.

[16] For a discussion of this program, see McCormmach 1970.

[17] Einstein 1907c discusses the defects of "our current electromechanical world-view," and looks forward to "a comprehensive world-view, corresponding to the relativity principle." See Holton 1981 for a discussion of Einstein's "Weltbild."

[18] For a more detailed later account by Einstein of the crisis in the foundations of physics initiated by Planck's studies of thermal radiation, see Einstein 1979, pp. 34–42. For his reaction to this crisis, see pp. 42–44. For Einstein's early study of blackbody theory, see Einstein 1989, editorial note, "Einstein's Early Work on the Quantum Hypothesis," pp. 134–148.

[19] In 1909 he noted "That it will only be a matter of a modification of our present theories, and not a complete abandonment of them" (Einstein 1909a, p. 192).

[20] See Einstein 1907c, pp. 372–373, for Einstein's comments on the range of applicability of Maxwell's equations. In the *Autobiographical Notes*, he commented: "The success of the theory of Brownian motion showed again conclusively that classical mechanics always led to trustworthy results whenever it was applied to motions in which the higher time derivatives of the velocity are negligible" (Einstein 1979, p. 46).

[21] For Besso's recommendation of Mach's books, see Einstein to Mileva Marić, 10 September 1899, Einstein 1987, Doc. 54, pp. 229–231, note 8. For evidence of a discussion with Besso, see Einstein to Marić, 4 April 1901, Einstein 1987, Doc. 96, pp. 284–286.

[22] Einstein probably read the third edition, Mach 1897, while a student. A copy was in his library until Helen Dukas gave it to Nathan Rosen. A more specific acknowledgment of Mach's role in the development of the theory of relativity is cited later in this section. For discussions of the relationship between Einstein and Mach, see Herneck 1966 and Holton 1968.

[23] Einstein speaks of studying the masters "with sacred enthusiasm."

[24] Einstein mentions his study of Kirchhoff, Helmholtz, and Hertz in his *Autobiographical Notes* (Einstein 1979, p. 14). For contemporary evidence of his reading of these authors, as well as of Boltzmann's *Gastheorie*, see Einstein 1987: Einstein to Marić, 10 August 1899 (Doc. 52), 10 September 1899 (Doc. 54), 1 August 1900 (Doc. 69), 13 September 1900 (Doc. 75), and 15 April 1901 (Doc. 101).

[25] For Einstein's discussions with Kleiner, see Einstein to Mileva Marić, 19 December 1901, Einstein 1987, Doc. 130, pp. 328–329. For a possible influence of Kleiner's views on Einstein see note 42 below.

[26] The earliest mention of the Academy is in a postcard of 30 November 1903 from Einstein to "Herr Dr. Conrad Habicht missing member of the A[cademy] of S[cience] [+] O[lympia]" (Einstein 1993a, Doc. 15, p. 24).

[27] For reminiscences of Einstein, which include a discussion of the Olympia Academy, see Solovine 1956, pp. v–xiii.

[28] I mention the works in the order in which Solovine lists them. He indicates that they read many other works on similar themes, in addition to a number of literary works. In a letter of 14 April 1952 to Carl Seelig (SzZE Bibliothek, Hs 304: 1006), Solovine includes Poincaré 1905 in an otherwise similar, but less complete list of the readings of the Olympia Academy

[29] Three editions came out within a few years: Mach 1900, Mach 1902, and Mach 1903.

[30] Solovine adds: "which Einstein already had gone through previously" (Solovine 1956, p. viii). Einstein probably read the third edition, Mach 1897, while he was a student. The Olympia Academy members may have used the fourth or fifth editions, Mach 1901, Mach 1904.

[31] Solovine notes: "Book III of Mill's Logic concerning induction also received our attention for a considerable period" (Solovine 1956, p. viii). The 8th ed., Mill 1872, was translated into German twice: Mill 1877, Mill 1884–1887.

[32] Solovine states: "For weeks we discussed the singularly sagacious critique of the notions of substance and of causality made by David Hume" (Solovine 1956, p. viii). For Einstein's comments on Hume see the text and note 38 below.

[33] Einstein presumably used one of the German translations: Spinoza 1887, Spinoza 1893.

[34] Helmholtz's scientific papers are published in three volumes: Helmholtz 1882, 1883, 1895 (the epistemological papers are included in Helmholtz 1883). His more popular writings are published in Helmholtz 1884. The fact that the Academy members read Riemann 1854 (see below) suggests that they also read Helmholtz's writings on the foundations of geometry, such as Helmholtz 1868.

[35] Clifford's essay was issued as a book in German, Clifford 1903, which Einstein presumably used.

[36] Einstein may have used the German edition, Poincaré 1904a, which contains extensive additional notes by Ferdinand Lindemann.

[37] In 1915, Einstein wrote to Schlick: "You have also correctly seen that this current of thought [positivism] had a great influence on my efforts, and indeed E. Mach and even more Hume, whose Treatise on Understanding I studied with fervor and admiration shortly before the discovery of the theory of relativity. It is very well possible that without these philosophical studies I would not have arrived at the solution" (Einstein to Moritz Schlick, 14 December 1915 (Einstein 1998, Doc. 165, p. 220). The reference to "Treatise on Understanding" is a conflation of the titles of two of Hume's works. Solovine's memoir suggests that it was the *Treatise on Human Nature* that Einstein read (see p. 7). See also Einstein to Michele Besso, 6 January 1948, and 6 March 1952, p. 7 (Speziali 1972, p.464). In the latter, Einstein stated that he read Hume "in a quite good German edition," presumably Hume 1895, a translation of the first part of Hume 1739.

[38] For the distinction between theories of principle and constructive theories, see Einstein 1919. He later reminisced about the origins of the special theory: "Gradually I despaired of the possibility of discovering the true laws by means of constructive efforts based on known facts. The longer and the more desperately I tried, the more I came to the conviction that only the discovery of a universal formal principle could lead us to assured results. The example I saw before me was thermodynamics" (Einstein 1979, p. 48). For several years after 1905, Einstein referred to the relativity principle, rather than to the theory of relativity.

[39] For a study of the physics of principles in Poincaré, and its antecedents, see Giedymin 1982, chap. 2, "The Physics of Principles and its Philosophy: Hamilton, Poincaré and Ramsey," pp. 42–89.

[40] The German edition of Violle's textbook, Violle 1892, which Einstein studied (see Einstein 1987, p. lxiv), characterizes physical principles as "originally hav[ing] developed from experience, but now occupy[ing] a much higher position; they are a type of postulate, the correctness of which is proved only by the continually perceived reliability of the conclusions to which they lead" (p. 90).

[41] Alfred Kleiner, with whom Einstein had been in contact since 1901, when he began to discuss physics with him (see Einstein to Mileva Marić, 19 December 1901, Einstein 1987, Vol. 1, Doc. 130), argued for the importance of such general principles as guides in the search for new theories, especially where commonly accepted constructive hypotheses encounter seemingly insurmountable obstacles (Kleiner had in mind primarily difficulties with mechanical models of the ether). In 1901 he wrote: "It must be emphasized that, in the extension of the sciences, we are directed to the aplication of general principles, if we want to . . . go beyond the sphere of the momentary state science has reached." He cited as examples of such principles: "The law of the constancy of matter, the principle of conservation of energy, the second law of thermodynamics, the ether hypothesis, etc." (Kleiner 1901, pp. 131–132).

[42] A decade later, Einstein elaborated this idea, introducing the concept of "constructive theories": "These attempt to construct a picture of more complex phenomena from a relatively simple underlying formalism. . . . When we say that we have succeeded in understanding a group of natural processes, we always mean by this that a constructive theory has been found, which embraces the processes in question" (Einstein 1919).

[43] "A satisfactory theory should in my opinion be so constructed that the electron appears as a solution, so that external fictions are not needed, in order not to have to accept that its electrical charges fly apart" (Einstein to Sommerfeld, 14 January 1908, Einstein 1993a, Doc. 73, pp. 86–89). This is presumably a reference to the Poincaré stresses.

[44] See Einstein 1911 for a discussion of these two principles.

[45] This attempt seems to represent Einstein's first step toward a field ontology.

[46] His analogy between relativity theory and thermodynamics as theories of principle may have prepared him to accept that, like the second law of thermodynamics, the original principle of relativity has its limits.

References

Ampère, André-Marie (1834). *Essai sur la philosophie des sciences ou exposition analytique d'une classification naturelle de toutes les connaissances humaines*. Paris: Bachelier.

Avenarius, Richard (1888). *Kritik der reinen Erfahrung*. Vol. 1. Leipzig: O.R. Reisland.

Avenarius, Richard (1890). *Kritik der reinen Erfahrung*. Vol. 2. Leipzig: O.R. Reisland.

Clifford, William Kingdon (1903). *Von der Natur der Dinge an sich*. Hans Kleinpeter, trans. and ed. Leipzig: Johann Ambrosius Barth.

Dedekind, Richard (1893). *Was sind und was sollen die Zahlen?* 2nd ed. Brauhnschweig: Friedrich Vieweg und Sohn.

Deltete, Robert John (1983). "The Energetics Controversy in Late Nineteenth-Century Germany: Helm, Ostwald and Their Critics." 2 vols. Ph.D. dissertation. Yale University.

Dukas, Helen, and Hoffman, Banesh, eds. (1979). *Albert Einstein, the Human Side: New Glimpses from his Archives*. Princeton: Princeton University Press.

Einstein, Albert (1902). "Kinetische Theorie des Wärmegleichgewichtes and des zweiten Hauptsatzes der Thermodynamik." *Annalen der Physik* 9: 417–433. Reprinted in Einstein 1989, Doc. 3, pp. 57–73.

— (1905a). "Über einen die Erzeugung und Verwandlung des Lichtes betreffende heuristischen Gesichtspunkt." *Annalen der Physik* 17: 132–148. Reprinted in Einstein 1989, Doc. 14, pp. 150–166.

— (1905b). "Zur Elektrodynamik Bewegter Körper." *Annalen der Physik* 17: 891–921. Reprinted in Einstein 1989, Doc. 23, pp. 276–306.

— (1906). "Zur Theorie der Lichterzeugung und Lichtabsorption." *Annalen der Physik* 20: 199–206. Reprinted in Einstein 1989, Doc. 34, pp. 350–357.
— (1907a). "Die Plancksche Theorie der Strahlung und die Theorie der Spezifischen Wärme." *Annalen der Physik 228*: 180–190. Reprinted in Einstein 1989, Doc. 38, pp. 379–389.
— (1907b). "Bemerkungen zu der Notiz von. Hrn. Paul Ehrenfest: 'Die Translation deformierbarer Elektronen und der Flächensatz." *Annalen der Physik* 23: 206– 208. Reprinted in Einstein 1989, Doc. 44, pp. 410–412.
— (1907c). "Über die vom Relativitätsprinzip geforderte Trägheit der Energie." *Annalen der Physik* 23: 371–384. Reprinted in Einstein 1989, Doc. 45, pp. 414– 427.
— (1907d). "Über das Relativitätsprinzip and die aus demselben gezogenen Folgerungen." *Jahrbuch der Radioaktivität and Elektronik* 4: 411–462. Reprinted in Einstein 1989, Doc. 47, pp. 433–484.
— (1909a). "Zum gegenwärtigen Stand des Strahlungsproblems." *Physikalische Zeitschrift* 10: 185–193. Reprinted in Einstein 1989, Doc. 56, pp. 542–550.
— (1909b). "Über die Entwickelung unserer Anschauungen über das Wesen und die Konstitution der Strahlung." *Deutsche Physikalische Gesellschaft. Verhandlungen* 11: 482–500. Reprinted in *Physikalische Zeitschrift* 10 (1909): 817–825 and in Einstein 1989, Doc. 60, pp. 564–582.
— (1911). L'état actuel du problème des chaleurs spécifique." In *La théorie du rayonnement et les quanta. Rapports et discussions de la réunion tenue à Bruxelles, du 30 octobre au 3 novembre 1911. Sous les auspices de M. E. Solvay*. P. Langevin and M. de Broglie, eds. Paris: Gauthier-Villars, pp. 407–435; German text is published in Einstein 1993b, Doc. 26, pp. 521–543.
— (1913). "Max Planck als Forscher." *Die Naturwissenschaften* 1: 1077–1079. Reprinted in Einstein 1995, Doc. 23, pp. 561–563.
— (1914). "Antrittsrede des Hrn. Einstein." *Königlich Preussische Akademie der Wissenschaften* (Berlin). *Sitzungsberichte*: 739–742. Reprinted in Einstein 1996, Doc. 3, pp. 20–23.
— (1915). "Theoretische Atomistik." In *Die Kultur der Gegenwart. Ihre Entwicklung und ihre Ziele*. Paul Hinneberg, ed. Part 3, sec. 3, vol. 1, *Physik*. Emil Warburg, ed. Leipzig: B.G. Teubner, pp. 251–263. Reprinted in Einstein 1995, Doc. 20, pp. 521–533.
— (1917). *Über die spezielle und die allgemeine Relativitätstheorie. (Gemeinverständlich)*. Braunschweig: Friedrich Vieweg und Sohn. Reprinted in Einstein 1996, Doc. 42, pp. 421–507.
— (1918). "Motiv des Forschens." Cited from the version published under the title: "Prinzipien der Forschung." In *Mein Weltbild*, Carl Seelig, ed. Zürich: Europa Verlag, 1981, pp. 107–110.
— (1919). "Time, Space and Gravitation." *Times* (London). 28 November, p. 13.
— (1921). *Geometrie und Erfahrung. Erweiterte Fassung des Festvortrages gehalten an der Preussischen Akademie der Wissenschaften zu Belin am 27. Januar 1921*. Berlin: Julius Springer. Cited from the reprint in: *Mein Weltbild*.

Carl Seelig, ed. Zürich: Europa Verlag, 1981, pp. 119–127.
— (1955). "Erinnerungen—Souvenirs." *Schweizerische Hochschulzeitung* 28 (*Sonderheft*): 143–153.
— (1979). *Autobiographical Notes. A Centennial Edition.* P.A. Schilpp, ed. La Salle, Ill.: Open Court, 1979.
— (1987). *The Collected Papers of Albert Einstein.* Vol. 1, *The Early Years, 1879–1902.* John Stachel, et al., eds. Princeton: Princeton University Press.
— (1989). *The Collected Papers of Albert Einstein.* Vol. 2, *The Swiss Years: Writings, 1900–1909.* John Stachel, et al., eds. Princeton: Princeton University Press.
— (1993a). *The Collected Papers of Albert Einstein.* Vol. 5, *The Swiss Years: Correspondence, 1902–1914.* Martin J. Klein, et al., eds. Princeton: Princeton University Press.
— (1993b). *The Collected Papers of Albert Einstein.* Vol. 3, *The Swiss Years: Writings, 1909–1911.* Martin J. Klein, et al., eds. Princeton: Princeton University Press.
— (1995). *The Collected Papers of Albert Einstein. Vol. 4, The Swiss Years: Writings, 1912–1914.* Martin J. Klein, et al., eds. Princeton: Princeton University Press.
— (1996). *The Collected Papers of Albert Einstein.* Vol. 6. *The Berlin Years: Writings, 1914–1917.* A. J. Kox, et al., eds. Princeton.: Princeton University Press.
— (1998). *The Collected Papers of Albert Einstein.* Vol. 8. *The Berlin Years: Correspondence, 1914–1918.* Robert Schulmann, et al., eds. Princeton: Princeton University Press.
Giedymin, Jerzy (1982). *Science and Convention: Essays on Henri Poincaré's Philosophy of Science and the Conventionalist Tradition.* Oxford and New York: Pergamon Press.
Helmholtz, Hermann von (1868). "Ueber die Thatsachen, die der Geometrie zum Grunde liegen." *Königliche Gesellschaft der Wissenschaften zu Göttingen. Nachrichten.* No. 9 (3 June): 193–221. Reprinted in Helmholtz 1883, pp. 618–639.
— (1882). *Wissenschaftliche Abhandlungen.* Vol. 1. Leipzig: Johann Ambrosius Barth.
— (1883). *Wissenschaftliche Abhandlungen.* Vol. 2. Leipzig: Johann Ambrosius Barth.
— (1884. *Vorträge und Reden.* 2 vols. Braunschweig: Friedrich Vieweg und Sohn.
— (1895). *Wissenschaftliche Abhandlungen.* Vol. 3. Leipzig: Johann Ambrosius Barth.
Herneck, Friedrich (1966). "Zwei Tondokumente Einsteins zur Relativitätstheorie." *Forschungen und Fortschritte* 40: 133–135.
Holton, Gerald (1968). "Mach, Einstein, and the Search for Reality," *Daedelus* 97: 636–673.
— (1981). "Einstein's Search for the *Weltbild*," *Proceedings of the American*

Philosophical Society 125: 1–15.
Hume, David (1739). *A Treatise of Human Nature: Being an Attempt to Introduce the Experimental Method of Reasoning into Moral Subjects*. Book 1, *Of the Understanding*. London: John Noon.
— (1895). *Ein Traktat über die menschliche Natur*. Part 1, *Über den Verstand*. E. Köttgen, trans., Theodor Lipps, ed. Hamburg: Leopold Voss.
Jungnickel, Christa, and McCormmach, Russell (1986). *Intellectual Mastery of Nature: Theoretical Physics from Ohm to Einstein*. Vol. 1, *The Torch of Mathematics 1800–1870*. Vol. 2, *The Now Mighty Theoretical Physics 1870–1925*. Chicago: University of Chicago Press.
Klein, Martin (1967). "Thermodynamics in Einstein's Thought." *Science* 157: 509–516.
— (1972). "Mechanical Explanation at the End of the Nineteenth Century." *Centaurus* 17: 58–82.
Kleiner, Alfred (1901). "Über die Wandlungen in den physikalischen Grundanschauungen." In *Verhandlungen der Schweizerischen Naturforschenden Gesellschaft bei ihrer Versammlung zu Zofingen den 4., 5. und 6. August 1901 (84. Jahresversammlung)*. Zofingen: P. Ringier, pp. 3–31.
Lorentz, Hendrik Antoon (1895). *Versuch einer Theorie der electrischen und optischen Erscheinungen in bewegter Körpern*. Leiden: E.J. Brill.
Mach, Ernst (1897). *Die Mechanik in ihrer Entwicklung. Historisch-kritisch dargestellt*. 3rd ed. Leipzig: F.A. Brockhaus.
— (1900). *Die Analyse der Empfindungen und das Verhältnis des Physischen zum Psychischen*. 2nd enl. ed. Jena: Gustav Fischer.
— (1901). *Die Mechanik in ihrer Entwicklung. Historisch-kritisch dargestellt*. 4th ed. Leipzig: F.A. Brockhaus.
— (1902). *Die Analyse der Empfindungen and das Verhältnis des Physischen zum Psychischen*. 3rd enl. ed. Jena: Gustav Fischer.
— (1903). *Die Analyse der Empfindungen and das Verhältnis des Physischen zum Psychischen*. 4th enl. ed. Jena: Gustav Fischer.
— (1904). *Die Mechanik in ihrer Entwicklung. Historisch-kritisch dargestellt*. 5th ed. Leipzig: F.A. Brockhaus.
McCormmach, Russell (1970). "H.A. Lorentz and the Electromagnetic View of Nature." *Isis* 61: 459–497.
Mill, John Stuart (1872). *A System of Logic Ratiocinative and Inductive: Being a Connected View of the Principles of Evidence and the Methods of Scientific Investigation*. 2 vols. 8th ed. London: Longmans, Green, Reader, and Dyer.
— (1877). *System der deductiven und inductiven Logik. Eine Darlegung der Principien wissenschaftlicher Forschung, insbesondere der Naturforschung*. 4th ed. J. Schiel, trans. Braunschweig: Friedrich Vieweg und Sohn.
— (1884–1887). *System der deductiven und inductiven Logik. Eine Darlegung der Principien wissenschaftlicher Forschung, insbesondere der Naturforschung*. 3 vols. 2nd ed. Theodor Gomperz, trans. Leipzig: Fues.
Minkowski, Hermann (1908). "Die Grundgleichungen für die elektromagnetischen

Vorgänge in bewegter Körpern." *Königliche Gesellschaft der Wissenschaften zu Göttingen. Mathematische Klasse. Nachrichten*: 53–111.

Nye, Mary Jo (1972). *Molecular Reality: A Perspective on the Scientific Work of Jean Perrin*. London: Macdonald.

Pearson, Karl (1900). *The Grammar of Science*. 2nd ed. London: Adam & Charles Black.

Planck, Max (1896). "Gegen die neuere Energetik." *Annalen der Physik und Chemie* 57: 72–78.

Poincaré, Henri (1902). *La science et l'hypothèse*. Paris: E. Flammarion.

— (1904a). *Wissenschaft und Hypothese*. Ferdinand and Lisbeth Lindemann, trans. Annotations by Ferdinand Lindemann. Leipzig: B.G. Teubner.

— (1904b). "L'état actuel et l'avenir de la physique mathématique." *Bulletin des sciences mathématique* 28: 302–324.

— (1905). *La valeur de la science*. Paris: E. Flammarion.

Riemann, Bernhard (1854). "Ueber die Hypothesen, welche der Geometrie zu Grunde liegen." *Königliche Gesellschaft der Wissenschaften und der Georg-Augusts-Universität* (Göttingen). *Mathematische Classe. Abhandlungen* 13 (1867): 133–152. Lecture, Göttingen, 10 June 1854.

Solovine, Maurice, ed. and trans. (1956). *Albert Einstein: Lettres à Maurice Solovine*. Paris: Gauthier-Villars.

Speziali, Pierre, ed. (1972). *Albert Einstein-Michele Besso. Correspondance, 1903–1955*. Paris: Hermann.

Spinoza, Baruch (1677). "Ethica ordine geometrico demonstrata.." In *Opera postuma*. Amsterdam: J. Rieuwertsz., pp. 1–264.

— (1887). *Die Ethik*. J. Stern, trans. Leipzig: Reclam.

— (1893). *Benedict von Spinoza's Ethik*. 5th ed. J.H. von Kirchmann, trans. Berlin: Philosophisch-Historischer Verlag.

Violle, Jules (1892). *Lehrbuch der Physik*. German edition by E. Gumlich et al. Part 1, *Mechanik*. Vol. 1, *Allgemeine Mechanik und Mechanik der festen Körper*. Berlin: Julius Springer.

The Mysteries and Wonders of Natural Science: Aaron Bernstein's *Naturwissenschaftliche Volksbücher* and the Adolescent Einstein

Frederick Gregory

1. Introduction

In his "Autobiographical Notes," Einstein tells us that an early religious phase around the age of twelve came to an abrupt end through the reading of popular scientific books (Einstein 1979, p. 2). Which books these were we learn from his sister's biographical sketch and from Max Talmey, the young Polish medical student who helped move the adolescent's focus beyond mathematics to philosophy and natural science. Maja Winteler-Einstein records that Talmey recommended to young Albert the *Kosmos* of Alexander von Humboldt, Ludwig Büchner's *Kraft und Stoff*, Aaron Bernstein's *Naturwissenschaftliche Volksbücher*, and other materials (Winteler-Einstein 1987, p. lxii). According to Talmey, the young Einstein was eager to discuss natural science, including the copies of Büchner and Bernstein that Talmey gave him (Talmey 1932, p. 162).[1]

There is no surprise in the mention either of von Humboldt's *Kosmos* or Büchner's *Kraft und Stoff*. William Langer reports that earlier in the century the *Kosmos* was read more than any other work except the Bible, while the sensation Büchner's book produced and its widespread availability in the late nineteenth century have been well documented (Langer 1969, p. 535; Gregory 1977, pp. 100–121). Few, however, are familiar with Aaron Bernstein, also known under the pen name A. Rebenstein. Depending on the edition, Bernstein's *Naturwissenschaftliche Volksbücher* ran to twenty-one volumes and, like Büchner's works, they were translated into several foreign languages. Although commonly known in the nineteenth century as first-rate popular-scientific accounts of developments in natural science, earning praise, for example, from Alexander

Einstein Studies, vol. 8: Einstein: The Formative Years, pp. 23–41.

von Humboldt himself (Mühsam 1940, p. 232), Bernstein's efforts have unfortunately been largely neglected since then. They deserve our attention, however, because they provide us with a means of assessing not only which of the amazing scientific results that were rapidly accumulating in the nineteenth century reached the reading public, but also how and in what form those results were presented to readers.

Where Einstein's early years are concerned, Bernstein's survey of the developments in natural science exposes us to an important part of the public's perception of natural science during the second half of the nineteenth century. As a boy, Einstein could not be immune to the manner in which the significance of natural science was understood by those around him. Popular writings on science played a major role in shaping the public's view of science; hence it is important to explore every avenue of entry available into the late nineteenth century German cultural milieu. Bernstein's books provide one such avenue. A measure of their significance is evident from Talmey's report of his meeting with Einstein in 1921. Talmey asked Einstein what he thought of a recent vilification of Bernstein's books as "sham science," and records this answer: "These are almost the very words of Professor Einstein's reply: Bernstein's work is a very good book even now, and at the time it was the best of its kind. It has exerted a very great influence on my whole development" (Talmey 1932, p. 163). There are, of course, several caveats that must be kept in mind with a source like Bernstein. Not only do we not know which edition Einstein read, we can also only surmise which of the topics treated would have interested him. As a result, the value of this series of materials has been judged at only the most general level. Thus, the editors of volume 1 of *The Collected Papers* are content to conclude merely that the materials Talmey gave Einstein "acquainted him with the mechanistic outlook in science" (Einstein 1987, pp. 5–6).[2]

It is also necessary at the outset to acknowledge an inevitable whiggishness in any treatment of Bernstein's importance to Einstein, for, since one cannot discuss everything covered in the 750 essays Bernstein composed, the selection of topics has been informed by later events in Einstein's life. One must presume, for example, that Bernstein's fascinating discussions of light and electricity are more relevant than is his treatment of nutritional science and health.

Such assumptions naturally make the historian uncomfortable since we simply do not know which of Bernstein's essays Einstein actually read. Even more problematical is the question of influence. Einstein himself tells us that popular scientific works affected his adolescent religious views. But could Bernstein's discussion of specific topics, for example, the nature of light, also have made an impression of significance on Einstein? No claims of this sort will be found below. The goal of this analysis is to describe what was contained in the essays Einstein had at his disposal and to allow readers to judge their importance for themselves. Nevertheless, it should be understood at the outset that through the selection of topics the deck has already been stacked to a certain extent.

2. A Biographical Sketch of Bernstein

Aaron David Bernstein possessed one of the most versatile minds of the nineteenth century. Born in Danzig in 1812, he was the son of a rabbi and was himself intended by his parents for the rabbinate. As a youth he received a thorough but exclusively religious education in the Talmud schools of Danzig and Fordon. We do not know the details of his rebellion against his parents' plans for his future. What we do know is that he left Danzig in 1832 at the age of twenty and went to Berlin, where he attempted to make up for his complete lack of a secular education through rigorous and exhaustive self-study. His first task was to improve his knowledge of German, but, judging from his later accomplishments, Bernstein also threw himself into the study of literature, political economy, history, and natural science.

Early on, Bernstein supported himself as an antiquarian bookseller, but it was as a professional writer that he was to leave his mark on German cultural life in the middle decades of the nineteenth century. His widely read stories, which were written in Judendeutsch, the German-Jewish dialect, have been called "forerunners of a literary genre that sentimentalized the Jewish lower middle class in small town ghettos" (Weltsch 1971, p. 682)

Bernstein hardly confined himself, however, to the writing of *Novelle*. In 1843 he cautiously entered the economic debate that centered on the policies of the Prussian Ministry of Finance. His defense of the Ministry, published in an anonymous pamphlet that was attributed by many to the Finance Minister himself, was among the earliest uses of statistics to shape public opinion (Nicolas 1955, p. 133; Wiernik 1906, p. 97). Throughout the 1840s Bernstein was active in the Jewish Reform Movement, an endeavor in which he was able to put his extensive religious schooling and his literary talents to good use. He was one of the committee appointed in 1845 to sketch out a program for the reform of Judaism, and as such he was among the principal authors of the call to organize the movement that appeared in Berlin's newspapers in April of 1845 (Wiernik 1906, p. 97). Along with Sigismund Stern, he wrote the prayer book used in the Berlin congregation, and he became editor of the monthly *Reform Zeitung: Organ für den Fortschritt im Judentum*, which commenced in 1847.

Bernstein's secular politics were also cut from a liberal mold. As a defender of democracy, he went to the barricades in the 1848 upheaval in Berlin. A year later he founded the *Urwählerzeitung*, a monthly journal upholding the need for political reform. Once the reaction to revolution had set in, a liberal political publication simply could no longer be tolerated. Bernstein was hauled into court in 1853 under the press law and sentenced to four months in prison. All of these experiences formed the basis for many political essays, which Bernstein later collected and published in three volumes bound as one (Bernstein 1882).

As varied as were these and other theological studies of higher criticism, it is yet another side of this fascinating man that interests us here. In 1853, the same year in which his political journal was suppressed, Bernstein embarked on yet another literary venture, this time a daily that soon had attained a wide circulation. For over twenty-five years Bernstein was the chief editorial writer for the *Berliner Volkszeitung*. It was here that his popular scientific essays first began to appear. In 1843 he had written a piece on the motion of the planets, but it was in his brief discussions of natural science, first in their serial form in the *Volkszeitung* and then in their several collected editions, that his promotion of the excitement and wonder of natural science was made available to Germans at mid-century.[3]

My treatment of the substance of Bernstein's essays is divided into three parts. A few initial observations about the impact that Bernstein's description of scientific achievements makes on the reader is followed by a discussion of Bernstein's handling of specifically selected topics. The analysis then concludes with an examination of the message contained in the essays about the significance of science for German thought, culture, and religion.

3. Bernstein's Impact as a Popular Science Writer

Anyone who picked up Bernstein's essays was bound soon to encounter the author's wholehearted enthusiasm for the awe-inspiring wonder of scientific investigation. "Natural scientists have considered and investigated things that strike any normal man as a fable," he wrote in an essay on how scientists had calculated the weight of the earth (Bernstein 1870, vol. 1, p. 8). The French astronomer Urbain Leverrier's successful prediction of the location of a new planet perturbing the orbit of Uranus was covered in a piece called "The Wonders of Astronomy," and elicited the following exclamation: "Praised be this science! Praised be the men who do it! And praised be the human mind, which sees more sharply than does the human eye!" (1870, vol. 16, p. 10).

At times one has the impression that Bernstein was aiming his message directly at young readers. In volume 16, for example, he invited his readers to join him for a fantasy journey into space. As with any other journey, he noted, travelers would require a suitcase, a ticket, and provisions for the trip. In the suitcase he proposed to carry ideas, the ticket would indicate the stops to be made, and for provisions he needed only a capacity for conversation. The means of travel would be neither horse, nor railroad, nor water; rather, the travelers would be born on the wings of a telegraphic signal (vol. 16, p. 11).

The entire trip lasted for thirty-six essays of volume 16 and the first eleven of volume 17. As the travelers left the earth for the moon, they looked back and reflected on the place that humans occupied in the larger scheme of things that lay ahead. Germany was no longer discernable, though Europe's outline could still be detected. As for humans, "these poor worms that crawl around on the lowest floor of the atmosphere," they were as good as non-existent (vol. 16, p. 14). On returning from his detailed guided tour of the solar system, Bernstein asked what effect the journey should have on our attitude. Should we be overwhelmed at our smallness in the universe, or should we be proud of the knowledge we have been able to acquire. Bernstein urged the middle road: We should give in neither to a haughty pride nor an oppressive humility, but we should pass on what we could to those who follow us (vol. 16, pp. 44, 48–49).

In addition to underscoring the pure excitement of scientific research, Bernstein's essays fulfilled a definite teaching function. It is clear from his style that Bernstein had assumed the role of an instructor. "The name of the fourth planet is Mars," he wrote in an essay explaining how light's power diminishes with distance from its source (vol. 1, p. 21). While he never wrote down to his audience, Bernstein communicated his lessons on science with homespun examples that would pique the interest of virtually anyone, to say nothing of a gifted adolescent like Einstein. Frequently Bernstein described in detail simple experiments that readers could do for themselves with materials readily available. Many of his essays consisted of carefully crafted reports of recent experiments made by scientists both inside and outside of Germany. Virtually all the great names of nineteenth-century natural science find their way into the narrative at some point or other.

Finally, Bernstein's readers were constantly reminded about the practicality of scientific research. Indeed, Bernstein chided his fellow Germans for not appreciating the practical benefits of research as much as did the English, French, and Americans. Reporting on the work of Carl Friedrich Gauss and Wilhelm Weber that made possible the detection of weak electrical currents needed to develop the early needle telegraph, Bernstein doubted whether the Berlin Academy would deem the application of Gauss and Weber's work to be worthy of celebration. "As it is in all things with us Germans, so it is here," he wrote in volume 4. Germans make many discoveries, he went on, but they never get to the people in practical form. Only when the English or Americans had made scientific discoveries important to the life of the people did the Germans belatedly recognize their merit (vol. 4, p. 53).

4. Specific Topics of Possible Interest to Einstein

When selecting the themes that might have caught the eye of the young Einstein, it is perhaps natural to inquire what Bernstein had to say about the subject of nature's forces. He approached the topic by a round-about route, inviting the reader to imagine what the world would look like if human beings possessed either one less or one more sense than the five we have. If everyone were born without sight, our only idea of distance beyond that obtained from our limbs would come through sound. The world would not be like that of the blind people among us, since sighted people see for them (vol 3, pp. 1–4). It would indeed be a different world from the one we know.

On the other hand, what if humans had six senses? Clearly we would add enormously to our knowledge of nature, although we could not specify at all what information the extra sense would convey. The point was, wrote Bernstein, that there are forces in nature that would remain hidden to us unless they were somehow brought to the attention of our existing senses. By raising the question of how our view of nature depends on the senses we happen to have, Bernstein introduced the theme of volumes 3, 4, and 5: "The Secret Forces of Nature."

Under this theme Bernstein first turned to the secret forces associated with matter. Bernstein explained that many of nature's hidden forces had remained essentially unknown for centuries. The most fundamental kind of force, attraction, was not identified until the time of Newton, even though, he observed with tongue in cheek, people prior to Newton's time *had* possessed five good senses. Bernstein's position was that by means of experiments we have been able to demonstrate the existence of attractive forces, but that natural scientists nevertheless admitted that the essence of attractive force remained a great secret (vol. 3, pp. 8–12).

As Bernstein proceeded to describe nature's secret forces it rapidly became clear not only that there must be several *different* forces of attraction, but also that nature contained fundamental forces of repulsion as well. After demonstrating that a bar of iron must consist of atoms separated by interparticulate spaces, Bernstein asked about the role of force on the atomic level. While gravitational attraction acted between atoms that were distant from one another, there was a different attractive force that acted between atoms of a given substance that were close together (vol. 3, pp.36–39, 68–72, 79–80). But were there only this cohesive force of attraction, bodies would collapse in upon themselves; hence there must also be a repulsive force between atoms that sets up an equilibrium with the attractive forces. Bernstein went on to explain the effect of heat on the interatomic attractive and repulsive forces. The addition of heat to a solid, he explained, reduced the attractive forces, with the result that a liquid was formed. Additional heat removed all attractive force of cohesion so that only repulsive forces remained active (vol. 3, pp. 28–32). Were it not for gravitational attraction, for example, the gaseous ball that the earth once had been would have diffused throughout the universe (vol. 3, pp. 36–39).

When Bernstein turned to other examples of hidden forces of attraction and repulsion, those associated with magnetism and electricity, he explained to his readers that here it was necessary to introduce a new entity into the scientific account: magnetic and electrical *Urstoffe*. Bernstein confessed that the use of the term *Stoff* here was out of the ordinary; usually we reserve the name *Stoff* for things that we can weigh. In this case scientists called the *Stoffe* magnetic and electric fluids, but one should understand, cautioned Bernstein, that the names were just words for unknown things whose true essence was hidden.

Bernstein described how experiments with magnets had led scientists to conclude that magnetic fluids were present in each atom of, for example, an iron bar, and that magnetic action consisted of a separation of two primary magnetic fluids to the poles of each atom. Young readers in particular might have noticed Bernstein's habit of pointing out work yet to be done by future scientists. With regard to magnetic force he observed, "Everybody feels that natural science is here just at the beginning of its scientific conquests and that there still remains much, extraordinarily much to do" (vol. 3, pp. 109–110).

Electricity was another of nature's attractive and repulsive forces. Its fluids were invisible to our five senses, and, as in the case of magnetism, the seat of its action lay in the atom. In spite of the enormous difference in the way electrical and magnetic fluids were transmitted, Bernstein noted that it seemed reasonable to suspect that both secret forces of nature could be treated as a single force that expressed itself differently in different circumstances (vol. 3, p. 134).

In addition to his explanation of the nature of electricity, Bernstein was careful to describe the difference between what he called frictional electricity and contact electricity (Galvanism). He explained the fundamental operation of electrical machines, declaring that electromagnetic force would replace steam power as soon as it became as cheap to produce as steam. His extensive description of the various stages of the development of the technology of the telegraph once again celebrated the practical genius of the American versions. He wrote: "In this sense the American telegraph is really American; i.e., practically set up" (vol. 3, pp. 146–147; vol. 4, pp. 43–44, 65). Bernstein himself invented a means for sending two telegraph messages simultaneously over a single wire, for which he received a patent in 1856 (Mühsam 1940, p. 232; Wiernik 1906, p. 98). His enthusiasm once again culminated in a declaration of exciting things to come, for scientists at present stood merely at the beginning of discoveries in electromagnetism. The future promised applications even for the household (vol. 4, pp. 43–44).

Chemistry provided Bernstein with yet another secret force. When iron combined with the oxygen of the air to produce rust a secret process was also involved. "There is here the greatest probability," wrote Bernstein, "that a secret force of attraction is here at work that performs such wonderful things" (vol. 5, pp. 3–4). Later Bernstein became bolder where the nature of this force was concerned. After explaining how the hypothesis of atoms was consistent with the laws of chemical combination, he turned to possible links between chemical and electrical force. Because electrical effects could be found wherever there were chemical processes, and because chemical effects could be produced from electrical currents, Bernstein was led to conclude that the chemical force he sought was really electrical force exhibiting chemical effects besides its own phenomena (vol. 5, pp. 99–100). Bernstein described the electrolysis of water and the electroplating of metals to demonstrate the link between chemistry and electricity. In a later discussion of chemistry he gave vent to his belief that the forces of nature would some day be shown to be somehow united. Perhaps, he speculated, chemical and electrical phenomena were both manifestations of a single, as yet unknown force rather than one being derivative from the other, for chemistry too was standing before the open gates of its field of inquiry (vol. 7, pp. 84–85).[4]

Periodically throughout the series Bernstein expressed his complete amazement at the speed with which some of nature's forces acted. The very first essay of volume 1, entitled "The Speed of Nature's Forces," began with the observation that, when one spoke of the speed with which light traveled through space, it was usually taken as either a myth or an exaggeration. But Bernstein delighted to explain in a second essay how Charles Wheatstone had approximated the speed of another of nature's fast-acting forces, electricity, by means of a rapidly rotating mirror, obtaining, he reported, a value of 60,000 German miles or 276,000 English miles per second. Such speeds were simply mind-boggling to him (vol. 1, pp. 3–7; cf. also vol. 4, pp. 22–25).[5]

Light was also the subject of discussion in three later volumes. In volume 8 there were eleven essays on "The Speed of Light," the first of which opened with Bernstein declaring incredulously: "Light travels 188,600 miles in one second!" (vol. 8, p. 124).[6] Bernstein then corrected the common understanding of the nature of human sight. We say in ordinary language that we see an object, he noted, but really we experience a sensation occasioned by the arrival of rays of light from the object (vol. 8, p. 127).

Since our sense knowledge, then, depended on rays of light, Bernstein introduced the idea of the *Postenlauf* of light. Drawing on his readers' experience of receiving news in a letter that, depending on how long it had been in the mail, was several days or even weeks old, Bernstein explained that our observation of nature also brought us old information. Where objects close by were concerned the difference was undetectable; even the sense impressions of the moon took only a little over a second to reach us. But with the sun, and especially with the stars, we were viewing historical events. For a star that was 10,000 light years away, this had the curious implication that the light arriving now had left the star at a date before the creation itself, provided, of course, that one dated the creation as many in older generations did. Since light from our past was just now reaching distant locations in space and leaving even them behind, Bernstein reached the conclusion that the present was eternal (vol. 8, pp. 148–155).

After explaining how Olaf Römer had used the eclipses of the satellites of Jupiter to ascertain that light took some eight minutes to travel from the earth to the sun, Bernstein observed that Römer could not determine a precise value for the speed of light because he did not have accurate knowledge of the earth–sun distance. According to Bernstein this precise calculation was left until modern times, when techniques for measuring what he called light's twin sister, electricity, were adapted by Armand

Fizeau and then Léon Foucault to arrive at the present value of 298,000 km per second (vol. 8, pp. 134–140; vol. 17, pp. 99–113).[7]

A young Einstein could also have read Bernstein's explanation of James Bradley's determination of the aberration of light. Bernstein's treatment ran as follows. Were one to fire a bullet through both walls of a moving train car, the two holes left would give the impression that the bullet had been fired from a direction at an angle to the car. This false impression is created, he explained, because the car was moving. In looking for evidence of parallax, the astronomer, Bradley, had realized that because we are on a moving earth, light entering a telescope would not reach the bottom of the tube unless one corrected for the earth's motion by tilting the telescope. This meant, in effect, that we see a star to be located in a place where it is not. The great significance of this discovery for Bernstein was that Bradley had determined the aberration to be the same for all light, regardless of how far away its source might be, which suggested that all light traveled at the same speed (vol. 8, pp. 141–147). From this Bernstein drew the following noteworthy conclusion:

> Since each kind of light proves to be of exactly the same speed, the law of the speed of light can well be called the most general of all of nature's laws, pointing to a single general cause that lights up the whole universe. So from the speed of light we have come to the conclusion that there must be a common cause of the transmission of light, and this opens to us the path to the nature of light. We intend to introduce our readers later to what science has investigated regarding this. (vol. 8, p. 159)

He came back to the issue in volume 19, where the question of the nature of light provided him with an occasion to investigate the unity of nature's forces. In earlier times, he explained, light was assumed to be a particle. But when it was found that two rays of light could cancel each other under certain circumstances, light began to be viewed as similar to sound-waves of a space-filling medium. The aether, as this medium was known, should theoretically affect things moving through it. Since no effect was detected, however, doubts have been expressed about the existence of the aether. But the degeneration of the orbit of a comet discovered in 1843 in France, Bernstein reported, "has made the existence of the aether completely indubitable" (vol. 19, p. 22).

In volume 18 Bernstein had clarified how natural scientists had come to the conclusion that there was only so much force [*Kraft*] in the universe, and that all of our work amounted only to the transformation of one kind of force into another. Of course he could not resist using the conservation of force to support one of his favorite conjectures, repeated throughout his essays, namely, that all of the forces of nature were perhaps simply different manifestations of one basic force (vol. 18, pp. 8–27). As early as volume 5, Bernstein had written that "the next significant level of natural science will be one in which the unity of force can be demonstrated" (vol. 5, p. 143; cf. also pp. 37, 41).

In his treatment of what in volume 18 he was calling the old theory of special fluids for light, heat, magnetism, and electricity, Bernstein explained that scientists had largely discarded such fluids as independent entities because of the reciprocity [*Wechselseitigkeit*] that forces manifested among each other. It was now undeniable that all so-called different fluids must come from a single source (vol. 18, p. 36).

Early in volume 19, Bernstein explained how aether waves beyond the violet end of the spectrum had been shown to affect photographic plates, thereby linking together optical and chemical phenomena. Given that wave lengths below the red end had been shown to be heat waves, Bernstein suggested that what united everything was the conception of a vibrating aether. When, for example, pressure, impact, or friction set an atom into vibration, then the aether around it also vibrated. Heat, therefore, was an activity state [*Tätigkeitszustand*] of the atom, and the so-called force of heat was simply the transmission of this activity state. The same could be said of light, chemical affinity, and electrical force, each of which resulted from a modification of the characteristic vibrations of atoms. The undulations of the aether corresponding to the atomic vibrations were perceived by our senses as the separate forces of nature.

But what about Newton's gravitational force? Was it somehow to be included among the forces that arose from the vibrations of the aether? Here again Bernstein permitted himself the luxury of speculation. Suppose that gravitational attraction did not act in an uninterrupted flow, but was impulsive, with a pause between moments of action. As long as the impulses were rapid, no mechanical laws would be affected. Sound, light, and electromagnetism were all impulsive in this sense, yet we perceive them to act continuously. Why not the same for gravity? (vol. 19, pp. 138–141).

Bernstein went on to suppose that, if the sun's gravitational force were exerted in this fashion, the complicated three-dimensional vibration patterns set up in the aether because of the sun's proper motion with respect to the aether would make the sun's gravitational force the origin of all of the aether's vibrations. Since these vibrations were perceived by us as the various forces present on earth, that would make the sun's gravitational action the one source of all the forces we know (vol. 19, pp. 143–144; see also pp. 110–111, 134–135). Bernstein thus satisfied his need to demonstrate how the unity of nature's forces might be possible.

5. Bernstein and the Broader Significance of Science

Bernstein's writings are also interesting for the view they reflected concerning the significance of natural science for philosophy and religion. In some ways they confirm the common impression we have of mid- to late-nineteenth century views of the authority of natural science. Over and over again, for example, Bernstein castigated philosophers for their inability to deal with nature properly. What one calls philosophy, he wrote in volume 7, had from the standpoint of the natural scientist sunk to an empty game of preconceived ideas. Bernstein illustrated his claim with the concept of vital force, which in earlier times had been subjected to philosophical speculation but which now had to withstand the scrutiny of modern chemical investigation. The study of philosophy had sunk to a curiosity since the discoveries of natural science had punished old whimsical lies (vol. 7, p. 85).

The nature philosophers of the first quarter of the century came in for especially harsh treatment. In his explanation in volume 3 of how humankind had gotten beyond the medieval notion that a stone fell out of its desire to return to the earth, Bernstein noted that we now know better, "not from philosophers of nature [*Naturphilosophen*], for they to this day play around with similar foolish ideas like those of the Middle Ages, but from natural scientists [*Naturforscher*]" (vol. 3, p. 41).[8] In other volumes Bernstein observed that science never had fallen into greater errors than when it had given in to the pure speculation of reason, and never had it gotten itself out of such problems better than by means of loyal observation of the world as it appeared before trying to answer the question of "what holds the world together in its innermost being." The saddest and most comical chronicle of errors, Bernstein concluded, was evident in what philosophers from Aristotle to Hegel had produced as philosophy of nature (vol. 5, pp. 143–144).[9]

Bernstein's attitude here regarding *Naturphilosophie*, an attitude also plainly evident in the works of Ludwig Büchner (Büchner 1855, pp. xxii–xiv), reflected a bias that made numerous figures in the scientific community of Bernstein's day blind to the fundamental contribution and continuing influence of romantic natural science in the later nineteenth century. In the recent collection of essays entitled *Romanticism and the Sciences* (Cunningham and Jardine 1990), more than one contributor has noted the debt owed to romantic philosophy of nature by both the ideology and institutions of the new natural science that wished to dissociate itself from its overly idealistic parent. New institutional arrangements for natural science such as the *Gesellschaft Deutscher Naturforscher und Ärzte* and the establishment of *Naturforschung* in the curriculum of university studies in Berlin served as models for the institutionalization of the new natural science. "Moreover," argue the editors in their introduction to the volume,

> the self image of the new "men of science" was to be largely continued by Romantic themes—scientific discovery as the work of genius, the pursuit of knowledge as a disinterested heroic quest, the scientist as actor in a dramatic history, the autonomy of a scientific elite. (Cunningham and Jardine 1990, p. 8)[10]

Each of these themes is vividly represented throughout Bernstein's writing.

In his own way Bernstein indirectly confirmed the work of modern historians of science who have recognized the significance of *Naturphilosophie* for the discovery of the conservation of energy. Bernstein did so by apologizing for the fact that the questions surrounding the conservation of force "sound philosophical, for since so-called *Naturphilosophie* has ruined everything to do with mind and knowing in the first quarter of our century, questions that sound philosophical stand to drive away dozens of natural philosophers" (Bernstein 1870, vol. 18, p. 24). After explaining Robert Mayer's experiments on the mechanical equivalent of heat and their significance, Bernstein once again defended himself: "What we say here is not just a philosophically speculative manner of speaking, but . . . a very practical real fact" (p. 26; cf. also vol. 5, p. 143).[11]

In more direct fashion Bernstein echoed the romantic insistence on nature's unity. Like Goethe, Bernstein felt that nature's unity was primal, and that the divisions we make in the organization of our perceptions do not represent truth as much as they do our own interests. From our youth we humans are used to viewing the whole world as if it existed only for us. Plants we do not eat or use we call weeds [*Unkraut*], regions where we cannot live, wilderness. We seek in all things the side that has a relation to us, and thereby forget that it is not the truth of nature but our self-love that gives us such a deviant judgment of the world outside us. "[It is] even more unfortunate that the cleverest of people proceed in exactly the same manner with the knowledge of nature" (vol. 5, p. 145).[12] Even our divisions of time into days, years, decades, and centuries, or natural science into mineralogy, chemistry, and physics did not exist in reality. They were "only a means of helping us." To nature there were no such divisions. A rock meant different things to a mineralogist, chemist, and physicist, but to nature it was just a unified stone (vol. 5, pp. 105–106).

And yet Bernstein was pulled in the other direction too, that is, away from the claim that our concepts only arbitrarily capture nature's unity. He clung fervently to the conviction that behind the complexities we observe in the appearances there lay a simplicity to which true representations of natural science conform. Nature was "very simple in what it creates, even if it appears to us highly mysterious and very complicated" (vol 5, p. 62). Furthermore, Bernstein's recognition of the self-interest of our categories did not prevent him from distinguishing natural science from speculative science by means of the former's sense of progress. In natural science "progress consists of the ever-growing precision with which the phenomena are observed and by which one is in a position to draw conclusions that come closer and closer to the truth" (vol. 17, p. 114).

The history of science represented to Bernstein the correction of past error and the accumulation of true results. While for Bernstein astronomy offered the best example of what he called a progress-science *[Fortschritts-Wissenschaft]*, he hoped a time of "unclouded truths in every intellectual area of human progress" was near (vol. 17, p. 117).[13] Near the end of the series Bernstein declared that his overview of natural science imparted confidence that the human investigative spirit was drawing closer to "the goal of true knowledge of the essence of natural phenomena" (vol. 19, p. 145). In the penultimate paragraph of the last volume he intoned that true science was "unending like the mind, and is caught in eternal progress like the very universe it strives to know" (vol. 19, p. 221; for similar sentiments, see also vol. 7, p. 90).

Bernstein celebrated the international spirit of science, denying that any single country could boast more than another about the human achievement. Since it had required Italians, French, Germans, English, and Americans united together to realize the triumphs of science, Bernstein concluded that it was the mind that united peoples of different lands, and that the time would come when the so-called great figures of the political realm would sink into the background as humankind fully realized that its true achievements lay in the peaceful progress of intellectual knowledge (vol. 19, p. 221).[14]

There remains the question of science and religion. How was religion portrayed by this rabbinically trained writer of prayer books? We have heard his enthusiastic hymn to the glories of natural science and we have encountered his profound distrust of traditional philosophy. But where did that leave Bernstein, who, in addition to his literary activities, was an active religious leader of his day?

It will come as no surprise that as a religious reformer Bernstein saw only advantage in the use of natural science to correct numerous specific traditional religious beliefs and even to oppose the older mentality in which everything was to be taken on faith. He was very clear that natural science could not acknowledge miracles because its very task was to reduce all phenomena to natural law. If, therefore, religion saw itself to be supernatural, then it could not expect to be aided by natural science (vol. 12, p. 50; see also vol. 3, pp. 122–123, 143).

Religion for Bernstein was one of humankind's intellectual inclinations [*geistige Neigungen*] along with the capacity for art and morality. Inclinations, according to Bernstein, represented undeniable human capacities that affected the way people lived their lives. Unlike instincts, however, inclinations required a free will (vol. 12, pp. 22–23). Religion was among the strongest inclinations, for those who gave in to it found that their lives were dominated by it.

Bernstein took considerable pains to defend the reality of the religious inclination in humans against those who attributed religion to erroneous understanding. One could not pretend that religious inclination was not there, wrote Bernstein, nor that it was either an intentional or unintentional delusion. The widespread existence of religion had to be acknowledged as a natural phenomenon with its own laws (vol. 12, pp. 49–51).

But what was the defining characteristic of the religious inclination if for some it expressed itself as a total embrace of supernaturalism while for others it did not?

> The innermost core of the religious inclination lies in the dim consciousness that dwells in humans that all nature, including the humans in it, is in no way an accidental game [*ein Spiel des Zufalls*], but a work of lawfulness, that there is a fundamental cause of all existence, and in it must lie the eternal source of all perishable phenomena. (vol. 12, p. 54)

According to Bernstein no one, even if rejecting religion as an illusion, could avoid assuming some natural principal as the source of all phenomena. The basic idea of all these presuppositions was and would always remain that something ruled the world, and that this unknown something was also active in us, guiding and conditioning us (vol 12, p. 55). Bernstein simply refused to acknowledge that it was possible for humans to be essentially unreligious creatures. Without religious inclination, he wrote, one could not conceive of an intellectually progressive and active humanity. Without it the whole existence of the world would appear to humans as an accident, and all thinking would be impossible, "for thinking involves consistency, and in accident there is no consistency." Everyone who wished to contemplate nature must dimly sense [*dunkel ahnen*] in advance that there was reason, or law, or spirit, or whatever else one wished to call it in nature. Without this, humans ceased thinking and no longer existed (vol 12, p. 55).[15]

Bernstein's treatment of the biological sciences has not been mentioned in this survey. Suffice it to say that while considerable space was devoted to such topics, there is, to my knowledge, no mention of the name of Charles Darwin anywhere in the series, and all mention of evolutionary theory is indirect. Bernstein gives evidence of a belief that not only the universe, but also human beings were very old indeed, but he did not take on evolution as a subject for one of his essays.[16] The reason seems clear. In his mind, chance, the phenomenon that the religious inclination could not tolerate at the heart of nature, was a central feature of Darwin's understanding of natural selection. *Evolution* seemed to present no essential problem either to his view of religion or his fervent belief in progress, but natural selection was another matter. Unlike many protestant theologians who embraced evolution while rejecting natural selection, however, Bernstein had identified himself too closely with natural scientists to fly in their face so easily by openly rejecting Darwin. Better simply to ignore the subject.

In light of the subject matter and the interpretations of Einstein's later work, it is evident that there is a great deal in Bernstein's popular works on natural science that is highly suggestive of Bernstein's influence. There is no evidence, however, on which to base a claim that Bernstein influenced Einstein in any way other than that which Einstein himself has indicated, namely, as the means by which he got over his adolescent religious phase. But in general it can be said that Bernstein's *Naturwissenschaftliche Volksbücher* provide us with an excellent vehicle by which we can understand how the ideas and results of natural science were communicated to the German public of Einstein's youth.

NOTES

[1] In a letter to Talmey of 21 January 1933, Einstein recorded his general approval of Talmey's *The Relativity Theory Simplified and the Formative Period of Its Inventor* (Talmey 1932). He did, however, offer several corrections where appropriate. Of the personal details in the book, Einstein corrected Talmey on one point only; namely, it was Einstein's uncle Jakob who had given him Heinrich Lübsen's calculus textbook (Lübsen 1869), not Talmey (1932, p. 164). Had Talmey not been the source of the Bernstein volumes, Einstein surely would have noted it,

especially since Talmey quotes Einstein as having reminded Talmey in 1921 that he had been the one who gave Bernstein's books to Einstein (1932, p. 163, n.). I am indebted to Professor John Stachel for bringing Einstein's letter to my attention.

[2] Lewis Pyenson is less generous, concluding that Bernstein's and Büchner's works were less significant for Einstein than Theodor Spieker's geometry textbook (Spieker 1890), which Talmey had also given to the young Einstein; see Pyenson 1985, p. 5.

[3] The first collection was entitled *Aus dem Reiche der Naturwissenschaft. Für Jedermann aus dem Volke* (Berlin: Besser); it appeared in twelve volumes between 1853 and 1857, with a second edition in 1858–1861. Beginning in the late 1860s several new editions appeared under the title *Naturwissenschaftliche Volksbücher* (Berlin: Franz Duncker). Issued in 20 volumes, then 21, the sets were bound in four or five separate books. It is unclear which edition of Bernstein's essays Einstein read, though Einstein's sister explicitly makes reference to the second title. Bernstein expanded the new editions as they appeared, both by including new materials and by revising the old. Individual volumes in the 1870 set, which I consulted, consisted of an average of 35 essays, each four to five pages in length. A copy of the titles of the individual volumes reveals that Bernstein's interests ranged as widely among the various disciplines within natural science as they did outside it.

[4] The lengthy exposition of nature's secret forces was brought to a close with a reference to the phenomena of heat and light. Up to this point, Bernstein wrote, he had been dealing with forces that were separable from matter. With heat and light things were different. Scientists could produce heat and light, describe how they were reflected, bent, etc., but had not arrived at a conclusion about what they were. Bernstein promised that he would return to these subjects in later volumes (cf. vol. 5, p. 140). Earlier in volume 5 he had discussed light briefly in connection with the effects of sunlight on chemical combination (vol. 5, p. 31).

[5] The enormous overestimation reported here would seem to indicate a misprint of 60,000 for 40,000 German miles; see, for example, note 6 below.

[6] The figure given is 41,000 German miles.

[7] Bernstein showed no awareness of the determination of the astronomical unit from the transits of Venus in the eighteenth century, even though he explained the transit phenomenon during his fantasy trip to Venus in volume 16, pp. 45–46.

[8] When later in the same volume Bernstein credited positive magnetic fluid with the desire to unite itself with negative, he gave no indication that he was aware he was contradicting himself; see vol. 3, p. 106.

[9] Hegel's dissertation on the planets was ridiculed both here and in volume 11; see vol. 11, p. 96.

[10] In the same volume Simon Schaffer adds his own summary of the transformation of natural philosophy into natural science: "Patterns of work such as the heroic privilege of discovery, the use of a disciplinary history as a means of legitimation of the division of labour in the sciences and the integration of laboratory teaching and lecture performance were all aspects of Romantic natural philosophy and its aftermath, the emergence of organized natural science" (Schaffer 1990, p.

94). Finally, Malcolm Nicolson's understanding of Alexander von Humboldt, who of course stood much closer to the Romantic Era than did either of the young Einstein's other two authors, emphasizes the universal, synthetic science portrayed in von Humboldt's writings, one that attempted to comprehend both the diversity and unity of nature (Nicolson 1990, pp. 171, 180). As we have seen from his handling of nature's forces, Bernstein shared much with the very tradition he criticized.

[11] Bernstein felt that if one was careful enough in observation one avoided the errors of philosophers; see vol. 5, p. 104.

[12] For Goethe's similar sentiments see, his "Toward a General Comparative Theory," where we read: "Why should [man] not ignore a plant which is useless to him and dismiss it as a weed since it really does not exist for him?" The weed, adds Goethe, is "a child sprung from all of nature, one as close to her heart as the wheat [man] tends so carefully and values so highly" (Goethe 1988, p. 53).

[13] On the exemplary role of astronomy, see vol. 17, pp. 115–116, and vol. 1, p. 27.

[14] In a rare political comment Bernstein argued that communism was "unnatural and untrue," because it ran counter to the natural inclination for wealth that was an undeniable characteristic of humans; see vol. 12, pp. 2–4.

[15] Bernstein here reminds one of many of the liberal protestant theologians of the second half of the nineteenth century, none of whom he gives indication of having read.

[16] For Bernstein's treatment of topics related to evolution, see especially vol. 8, pp. 9–10, 41–43, 61–63, 69–71, 94–106, 118–123, and vol. 13, pp. 143–146.

References

Bernstein, Aaron (1870). *Naturwissenschaftliche Volksbücher*. Berlin: Franz Duncker.

—(1882). *Revolutions- und Reaktionsgeschichte Preussens und Deutschlands von den Märztagen bis zur neuesten Zeit*. Berlin: Wortmann.

Büchner, Ludwig (1855). *Kraft und Stoff*. Frankfurt am Main: Meidinger.

Cunningham, Andrew and Jardine, Nicholas, eds. (1990). *Romanticism and the Sciences*. Cambridge, UK: Cambridge University Press.

Einstein, Albert (1979). *Autobiographical Notes*. Paul Arthur Schilpp, ed. and trans. LaSalle, Illinois and Chicago: Open Court.

— (1987). *The Collected Papers of Albert Einstein*. Vol. 1, *The Early Years, 1879–1902*. Stachel, John, et al., eds. Princeton: Princeton University Press.

Goethe, Johann Wolfgang von (1988). "Toward a General Comparative Theory." In *Goethe*. Vol. 12, *Scientific Studies*. Douglas Miller, ed. and trans. New York: Suhrkamp, pp. 53–56.

Gregory, Frederick (1977). *Scientific Materialism in Nineteenth-Century Germany*. Dordrecht: D. Reidel.

Langer, William (1969). *Political and Social Upheaval, 1832–1852.* New York: Harper & Row.
Lübsen, Heinrich Borchert (1869). *Einleitung in die Infinitesimal-Rechnung (Differential- und Integral-Rechnung). Zum Selbstunterricht. Mit Rücksicht auf das Nothwendigste und Wichtigste*, 4th ed. Leipzig: Friedrich Brandstetter.
Mühsam, Hans (1940). "Aaron Bernstein." In *The Universal Jewish Encyclopedia*, vol. 2. New York: Universal Jewish Encyclopedia, pp. 232–233.
Nicolas, Marcel (1955). "Bernstein. Aaron David." In *Neue Deutsche Biographie*, vol. 2. Berlin: Duncker & Humblot, p. 133.
Nicolson, Malcom (1990). "Alexander von Humboldt and the Geography of Vegetation." In Cunningham and Jardine 1990, pp. 169–185.
Pyenson, Lewis (1985). *The Young Einstein: The Advent of Relativity.* Boston: Adam Hilger.
Schaffer, Simon (1990). "Genius in Romantic Natural Philosophy." In Cunningham and Jardine 1990, pp. 82–98.
Spieker, Theodor (1890). *Lehrbuch der ebenen Geometrie. Mit Übungsaufgaben für höhere Lehranstalten,* 19th impr. ed. Potsdam: Aug. Stein.
Talmey, Max (1932). *The Relativity Theory Simplified and the Formative Period of Its Inventor.* New York: Falcon Press.
Weltsch, Robert (1971). "Bernstein, Aron David." In *Encyclopædia Judaica*, vol. 4. New York: Macmillan, pp. 682–683.
Wiernik, Peter (1906). "Bernstein, Aaron (David)." In *The Jewish Encyclopedia*, vol. 3. New York: Funk and Wagnalls, pp. 97–98.
Winteler-Einstein, Maja (1987). "Albert Einstein–Lebensbild." In Einstein, 1987, pp. xlviii–lxvi.

The Young Einstein's Physics Education: H.F. Weber, Hermann von Helmholtz, and the Zurich Polytechnic Physics Institute

David Cahan

1. Introduction

In October 1897, Albert Einstein, then eighteen years old and at the start of his third semester at the Eidgenössische Polytechnische Schule in Zurich, enrolled to hear the physics lectures of Heinrich Friedrich Weber, professor of mathematical and technical physics and director of the physics institute at the school. Einstein's enrollment in Weber's course was an essential step towards achieving his goal of obtaining a diploma to teach high school physics and mathematics. Weber and his institute had much to offer in this regard, and, in a roundabout way, perhaps even something to offer towards Einstein's second, more ambitious and less pragmatic goal: to become a theoretical physicist.

During his years at the Zurich Polytechnic (1896–1900), Einstein sought to achieve his two goals through a tripartite physics education, with each part largely but not completely independent of the other. The first part consisted in the study of mathematics and mathematical physics. The Polytechnic provided superb resources in these fields—as the mere mention of the names of Adolf Hurwitz and Hermann Minkowski should indicate—and Einstein registered for a variety of courses in mathematics and mathematical physics (Einstein 1987, pp. 45–50, 362–369). Yet during his student years Einstein found most of these courses irrelevant to his physical interests, and in at least several of them, his participation was at best minimal. He apparently thought that the calculus and not much beyond it was all that he needed to pursue physics.

Einstein Studies, vol. 8: Einstein: The Formative Years, pp. 43–82.

After 1907, as he sought to fashion a general theory of relativity, and later still, as he sought to create a unified field theory, he came to regret his earlier, cavalier attitudes towards the importance of advanced mathematics for physics and towards the Polytechnic's offerings (Einstein 1949, p. 15; Einstein 1956, p. 11). The second part of Einstein's physics education consisted in the autodidactic study of the writings of the masters of contemporary theoretical physics—above all, Hermann von Helmholtz, Gustav Robert Kirchhoff, Heinrich Hertz, Ludwig Boltzmann, Ernst Mach, and, through the works of Paul Drude or August Föppl, James Clerk Maxwell. Einstein's autodidacticism was due in part to the shortcomings of the third part of his physics education, namely, attending Weber's lectures and working in his laboratory.

To the extent that Einstein can be said to have had formal training in physics and to have had a physics teacher, then it was in the courses that he registered for with Weber between 1897 and 1900.[1] Einstein's early association with Weber, his initial enthusiasm for and subsequent disappointments with Weber as a teacher, and, finally, his pleasure at working in Weber's laboratory—these and a few other related points have been mentioned *en passant* either by Einstein himself or by any number of Einstein scholars. That Weber and his institute to some degree and in some sense influenced Einstein is clear enough; yet the precise nature and extent of that influence remains to be delineated. Moreover, the recent publication of Einstein's extensive, carefully worked-up lecture notes from his first physics course under Weber (Einstein 1987, pp. 63–210) gives additional impetus to explore the general influence—both positive and negative—of Weber and his institute on Einstein, as well as the particular influence of Weber's work and style for understanding the scientific development of the young Einstein.

Yet to understand that influence fully—or at least as fully as the documentation seems to allow—requires an understanding of Weber himself, of his own professional development, of his research orientation and accomplishments, of the physics institute that he created at the Polytechnic in the late 1880s, and, finally, of his pedagogical gifts and administrative style. Moreover, as this essay shall further argue, due attention must also be given to the influence of Helmholtz on Weber and on Weber's associate at the Polytechnic's physics institute, Johannes Pernet. Much of this essay (Sections 2 and 3) is thus devoted to Weber and his institute: to relating his strengths and weaknesses as a physics researcher and teacher, and to re-creating the general atmosphere of his institute. Yet the ultimate purpose of this background analysis of Weber and his institute is to deepen appreciation of the second and third parts of

Einstein's physics education, his autodidactic study of theoretical physics and the influence of hearing Weber's lectures and working in his laboratory (Section 4).

Figure 1. Heinrich Friedrich Weber
Courtesy of the ETH Bibliothek, Zurich

2. H.F. Weber: The Helmholtzian and Precision-Measuring Background

Heinrich Friedrich Weber (1843–1912) was born in the town of Magdala, near Weimar, son of a merchant. Around 1861 he entered the University of Jena, where Ernst Abbe became the first of two physicists who decisively influenced his career (Weiss 1912, pp. 44–45).[2] In the early 1860s, Jena was anything but a scientific center. To name but one of its many short-comings, it lacked a good mathematician, which led Weber himself to try to become one. Weber soon discovered, however, that he lacked sufficient mathematical talent, and so he abandoned mathematics entirely (Weiss 1912, p. 44).

Returning to the fold of physics, Weber found in Abbe a young and dynamic scientist, one who successfully focused much of his research efforts on re-thinking optical theory so as to help create new or improved precision optical instrumentation, and one who successfully focused much of his teaching efforts on providing his students with good precision instruments and good laboratories (Auerbach 1922). During the first-half of the 1860s, when Weber studied at Jena, the physics institute there was one of the smallest and most poorly equipped of German physics institutes (Auerbach 1922, pp. 37–38, 46–48, 77–81; Cahan 1985, p. 13). Nonetheless, Abbe stressed physical laboratory work as much as his limited facilities would allow, and during the 1860s and beyond, he sought constantly to improve Jena's meager physics facilities, until in 1884, he finally managed to have an entirely new building constructed for physics (Cahan 1985, pp. 17, 22–24). Furthermore, Weber also found in Abbe a physicist who sought to unite science and technology, to turn scientific knowledge to practical use (Weiss 1912, p. 44).

That unification and transformation found their fullest expression during the 1870s and beyond, when Abbe played the central role in developing the Zeiss Optical Works in Jena (Auerbach 1922). Abbe thus not only instructed Weber in the lecture hall and laboratory, he also served as a role model for him in several other ways: Through his emphasis on the importance of laboratory work in general and precision instrumentation in particular; through his view that science should be closely related to practical life; and through his embodiment of the idea that a single individual could accomplish much in life. This last point, according to Pierre Weiss, Weber's obituarist and successor at the Eidgenössische Technische Hochschule, as the Zurich Polytechnic came to be called after 1911, was "the mainspring of his [Weber's] life, the source of his most beautiful successes" (Weiss 1912, p. 44).

Weber received his doctorate under Abbe in 1865 with a dissertation on the theory of light diffraction. He spent the second half of the 1860s as a private tutor in Pforzheim, publishing only one article during this period. However, his scientific research was apparently not quite as moribund as this fact might suggest. For Pforzheim was close to both the University of Heidelberg, where Weber came into contact with Kirchhoff, one of the leading theoretical physicists of the day, and to the Polytechnische Schule in Karlsruhe, where in 1870 he became Gustav Wiedemann's assistant (Weiss 1912, p. 45). At the same time, Weber also managed to meet the professor of physiology at Heidelberg, Hermann von Helmholtz, perhaps through Kirchhoff or Wiedemann, since both of them were close friends of Helmholtz's. Helmholtz had devoted much of the 1850s and 1860s to physiological optics and acoustics, in particular to problems in color theory and perception. By the late 1860s, however, he was devoting most of his time to his first love, physics, in particular to clarifying rival candidate laws of electrodynamics; and during the 1880s and early 1890s he turned to clarifying the mechanical foundations of thermodynamics and, through the use of the principle of least action, the foundations of mechanics itself. When Helmholtz left Heidelberg in 1871 to accept the call as professor of physics at the University of Berlin, he took Weber along as his first assistant. Helmholtz now became the second formative, and decidedly primary, influence on Weber's career.

During his three years as Helmholtz's assistant in Berlin (1871–1874), Weber helped Helmholtz set up and equip the Berlin laboratory and also helped him direct the student laboratories. The lessons that Abbe had first taught him about the importance of good laboratories and instrumentation were now reinforced by Helmholtz. Indeed, Helmholtz's laboratory was the first substantial, and certainly the first major laboratory in which Weber worked. It gave Weber the means to conduct both laboratory instruction and research. To the earlier interest in optics that he had inherited from Abbe, he now, under Helmholtz, added the fields of heat and electricity as his principal areas of research.[3]

In 1872 and 1875, Weber published two noteworthy papers in the *Annalen der Physik* on determining the specific heats of carbon, boron, and silicon at various temperatures (Weber 1872, 1875). Working in Helmholtz's refurbished Berlin laboratory, Weber measured the specific heats of these three elements and showed them to be noticeably smaller at low temperatures than predicted by the law of Dulong and Petit; moreover, he found that, with an increase in temperature, their specific heats increased extraordinarily rapidly. Only when he increased the temperature beyond 1,000°C did the specific heats again follow the predictions of Dulong and Petit. For nearly thirty years Weber's empirical findings remained an anomaly until, as we shall see, one of his students presented a new explanation (Weiss 1912, pp. 49–50; Pais 1982, pp. 391–392).

In 1874, Weber left Berlin and Helmholtz to become the professor of physics and mathematics at the undistinguished Königliche Württembergische Akademie in Hohenheim. During one of his classes, he suddenly found among his listeners an older gentleman. When the class ended, the man introduced himself as Karl Kappeler, who, from 1857 to 1888, was the Schulratspräsident of the Zurich Polytechnic. Then and there, Kappeler offered Weber the professorship of mathematical and technical physics in

Zurich (Weiss 1912, p.45). It was almost certainly on Helmholtz's recommendation that Kappeler had come to Hohenheim to see Weber, who began teaching in Zurich in 1875. For when Kappeler died in 1888, Weber felt compelled to report his death immediately to Helmholtz,

> because in this moment of such a deeply moving event I vividly feel what extraordinarily great benevolence and what extraordinary help for my efforts Herr Kappeler has always shown and because I owe to your kind benevolence alone the good fortune to have stood throughout the years in close contact with this outstanding and, in every sense, excellent man. I am grateful to you from the bottom of my heart forever. (Weber to Helmholtz, 21 October 1888, Helmholtz Nachlass, hereafter HHN; see the Acknowledgments)

As will soon become clear, Weber had still other reasons to be thankful to Kappeler and Helmholtz.

Following his work in Helmholtz's Berlin laboratory on the anomalous temperature dependence of specific heats, Weber published some twenty articles in the general fields of optics, heat, and electrical measurement, while working in his own laboratory in Zurich. His articles on electrical measurement included work on absolute precision electromagnetic and thermal measurements, on induction processes in telephony, on the absolute value of the Siemens mercury unit of resistance, on self-induction in metal conductors and bifilar wire rolls, and on the theory of Wheatstone bridges (Weber 1877, 1878, 1884, 1886a, 1886b, 1887a). During the last-third of the nineteenth century, Weber became a noteworthy, if not leading, measuring physicist. Like J.W. Strutt (Lord Rayleigh), William Thomson (Lord Kelvin), and Silvanus Thomson in Britain, and like Friedrich Kohlrausch and Helmholtz in Germany, Weber's work in metrology in general and electrical standards in particular helped bridge the gap between the metrological and instrumentational needs of academic physics laboratories, on the one hand, and those of the burgeoning electrical industry, on the other (Weiss 1912, pp. 48–49).

Along with his work on specific heats and electrical metrology, Weber published on thermal and electrical conductivity in metals, on liquid conductivity, on light emission, and on radiation. These four publications by Weber merit special mention for an appreciation of his general importance to physics in the 1870s and 1880s, the influence of Helmholtz on his career, and, in turn, Weber's influence on Einstein.

First, in the spring of 1880, Weber wrote a manuscript on the relation between thermal and electrical conductivity in metals. He sent the manuscript to Helmholtz, asking him to arrange for its appearance in the

Monatsberichte of the Berlin Akademie der Wissenschaften, which, in the event, Helmholtz did (Weber to Helmholtz, 26 April and 25 May 1880, both in HHN; Weber 1881). By mid-summer of that year, to turn to the second publication, Weber and Helmholtz were again in correspondence, this time to discuss their similar interests in the conductivity of liquids. Since at least 1879 Helmholtz had been concerned with problems of electrolytic conductivity. He now sent Weber a copy of his recently published paper on the subject (Helmholtz 1880). Weber responded that he too had just published on a very similar subject, namely, heat conductivity in liquids (Weber 1879). He wrote Helmholtz:

> Since the year 1877 I had sought to find an empirical unit for electromotive force that satisfies all demands and, further, to determine as exactly as possible the absolute value of this empirical unit according to different methods.

This work, Weber explained, concerned Daniell cells as well as "a series of theoretical considerations on hydroelectromotive forces and galvanic polarization, etc." The similarities to Helmholtz's work went beyond the common and typical use of Daniell cells for electrolytic studies, however. For Weber further reported that his theoretical results were *the same* as Helmholtz's (cf. Helmholtz 1880, pp. 295–301), and that, apparently unbeknownst to Helmholtz, he had published his results in May 1879. Weber continued:

> Immediately upon receipt of your article I thought it my duty to let you know of this circumstance, and I beseech you with the greatest respect, most honored Herr Privy Councilor, to consider this communication as an open expression of one of your students for whom you have opened the scientific paths in every direction and who will forever remain in grateful respect to you. (Weber to Helmholtz, 16 July 1880, HHN)

Here, then, is the thirty-six-year-old professor of mathematical and technical physics at the Zurich Polytechnic, a man who within a decade will be giving instruction in physics to the young Albert Einstein, explicitly acknowledging Helmholtz as his teacher and scientific pathfinder.

The third and fourth publications appeared in 1887 and 1888, respectively. In 1887, Weber, again thanks to Helmholtz, published in the Berlin Akademie's *Sitzungsberichte* and in the *Annalen der Physik* on light emission in incandescent solid bodies. Here he reported on his determination of the lowest temperature at which incandescence in such bodies could be detected (Weber to Helmholtz, 10 July 1887, HHN; Weber 1887b). He showed that, if the observer rested his eyes in complete darkness prior to making his observations, then emission could be detected at a much lower temperature—around 400°C—than that found by John William Draper, namely, around 525°C. Moreover, Weber denoted the "color" seen within this temperature range as "ghost gray" ("Gespenstergrau" or "Dünsternebelgrau"), since the eye cannot perceive color in this range.

This sort of study—namely, high-temperature precision measurements—was precisely the sort that Helmholtz would soon have several staff members undertake at his new Physikalisch-Technische Reichsanstalt, which opened its doors in 1887–1888 (Cahan 1989). Indeed, Helmholtz himself had been one of the leading students of the whole topic of human color perception (Kremer 1993). Then in 1888, Weber published what was arguably his premier piece of research: a study of the radiation spectrum of solid bodies, one which appeared, again thanks to Helmholtz, in the Berlin Akademie's *Sitzungsberichte* (Weber 1888). Here Weber challenged a formula proposed by Albert Michelson, another Helmholtz student, that was used to cover a set of temperature measurements, presenting, instead, his own set of measurements and a semi-empirical formula. While hardly definitive, his work nonetheless placed him in the forefront of research aimed at understanding the radiation spectrum of solid bodies. Its distinguishing feature was precision measurement: he presented an empirically-based spectral-energy distribution curve.

Weber's work came to be regarded as an important early contribution to the search for an energy distribution law for blackbody radiation, in particular serving as a starting point for Wilhelm Wien's displacement law, which, in turn, was crucial for Max Planck's work (Kangro 1970, pp. 3, 4, 38–41). Indeed, during the 1890s several investigators at the Reichsanstalt (including Wien) would be involved in such experimental and theoretical work (Cahan 1989, pp. 143–157). Yet Weber's interest here, like the interests of his Reichsanstalt colleagues, transcended pure science: he aimed to understand the properties of electric lighting by establishing a scientifically-grounded radiation formula (Weber 1888, esp. pp. 934–936, 957; Cahan 1989, pp. 143–157).

In sum, Weber's research in optics, heat, and electricity was essentially of a piece. In all three fields he used or developed empirically-based laws to cover, or low-level theory to explain, precisely measured empirical findings. Moreover, wherever possible, he sought to turn his results to practical use for sources of illumination, telephony, and the establishment of electrical standards. As Section 3 will show, that attitude of relating scientific and technical concerns was only fitting for the director of the Zurich Polytechnic Physics Institute.

3. The Zurich Polytechnic Physics Institute: Electrotechnology and Precision Measurement

With the publication of his study on radiation in 1888, Weber essentially ceased doing scientific research; with one partial exception (Weber 1897), his publications thereafter were limited to technical reports. Between 1883 and 1902 he published six articles on highly practical, technological subjects, including studies of the transmission of electrical energy between various cities, reports on the use of alternating current systems in electrical railroads, and a report on a Swiss federal law concerning low- and high-voltage equipment. More pertinent to our present purposes is the fact that, during Einstein's years at the Polytechnic, Weber published only one scientific paper, and that on a topic closely related to electrotechnological concerns: alternating currents (Weber 1897; cf. Weiss 1912, pp. 52–53). Moreover, Weber participated in electrical congresses at Vienna (1883), Paris (1889), Frankfurt am Main (1891), Paris (1900), and London (1908) (Weiss 1912, p. 49). He became the representative of Swiss electrical metrology, just as Thomson was for Britain and Helmholtz for Germany.

Weber's retreat from scientific research coincided with—and was most likely caused by—his efforts to construct, equip, and direct a new physics institute for the Polytechnic. From the narrow professional point of view, this was the outstanding event of Weber's career (Weiss 1912, p. 46). He undertook these efforts at a moment when new institutes for physics were being established throughout the German-speaking academic world (Cahan 1985), and when new institutes, chairs, and programs for electrotechnology were also being established throughout Europe and North America. Weber was determined that his new institute would be as devoted to electrotechnology as to physics per se (Weiss 1912, p. 47).

These changes in his professional orientation and activities are all the more striking, and significant for Einstein's education because they occurred in the very years (the late 1880s) in which Hertz began to publish his epochal results on electric waves and his confirmation of Maxwell's theory of electromagnetism. To put it differently, at the very moment when Europe's leading physicists came to accept the first significant change in the foundations of their discipline since the formulation of the laws of thermodynamics at mid-century, Weber entered into intellectual stasis. By 1890 he had apparently lost interest in and contact with the forefront of research in physics, at the very least in the field of electromagnetism. His understanding of electromagnetic phenomena and laws remained that of a previous generation. From the late 1880s, his preoccupation with building his new institute and with his practical or technical studies apparently left him insufficient time and energy to become adequately acquainted with Maxwell's results and their implications for everything from the foundations of physics to long-distance electrical communication systems.[4]

The central purpose of the mathematical and natural scientific departments at the Polytechnic was to train teachers of mathematics and natural science (Oechsli 1905, p. 325). Between 1895 and 1905, for example, most graduates of the Polytechnic's natural science departments became secondary school teachers or worked in private industry; only a handful took up academic careers or did any sort of research or testing work for the state (Frey-Wyßling and Häusermann 1955, p. 59). Accordingly, Weber's institute had a twofold purpose: to teach "scientific physics," as he put it, on the one hand, and, on the other, "to make possible the acquisition of fundamental knowledge of the most important parts of applied physics for contemporary technology through the close connection of instruction in the lecture hall and an introduction to intensive practical work in the physical laboratory" (Schweizerischer Bundesrat 1889, p. 52).

Both lectures and laboratory work in Weber's institute had a tripartite division. In the lecture hall, Weber and his associates offered lectures in general experimental physics as well as in specialized fields of physics, such as heat and electrical theory; in mathematical physics; and in applied physics, including topics like applications of electricity to technology, principles of electrotechnology, theoretical understanding of dynamos, accumulators, transformers, telegraphy, and so on. The laboratory work, for its part, sought to relate "the elementary physical exercises," "scientific work," and "the study of electrotechnology."

Hence there were three types of laboratories: First came the introductory physical laboratory where students worked five hours weekly in order to strengthen their understanding of physical laws and processes, to become acquainted with measuring instrumentation (not least for its importance in understanding physical laws and processes), and, finally, "to acquire for oneself every exercise of physical thought, every dexterity of the hands and every skill of the observing senses that are so absolutely necessary for more comprehensive and deeper studies in scientific and technical physics."

The second set of laboratories was reserved for the professional staff and "for the scientific training of advanced students who seek fundamental and comprehensive training as physicists." Advanced students worked in these laboratories for two years: "the first year, with 12 hours weekly, is devoted to exercises in exact scientific works in general, the second, with

24 hours weekly, is devoted to pursuing a specific scientific project in the most independent way possible."

The third set of laboratories was devoted to the study of electrotechnology, and here students worked twelve or twenty-fours hours weekly in order "to learn thoroughly the most important areas in the theory of electricity and the technical applications of electricity by means of their own practical work in a systematically ordered course of electrical measurements" (Schweizerischer Bundesrat 1889, pp. 52–54).

During his first fifteen years as director of the Polytechnic's physics institute (1875–90), Weber had to make do with the outdated, "wretched," and spatially inadequate laboratories that were part of the school's main building. These laboratories reportedly did not allow Weber and his associates to keep pace with modern research, a state of affairs "which makes the execution of some studies extremely troublesome and laborious, and the more delicate scientific investigations almost impossible" (Schweizerischer Bundesrat 1889, pp. 52–54). During the second half of the 1870s, the efforts of the Polytechnic's leaders to expand the school and undertake new construction, in particular to provide new physics laboratories and equipment, were repulsed by the canton of Zurich, which did not want to assume the necessary financial burden. By the 1880s, however, the increased, pressing need to give more and better laboratory instruction to students of chemical technology and mechanical engineering, as well as to science teachers could no longer be postponed. Thus, in 1883, the Polytechnic authorities—Kappeler and the mathematician Carl Friedrich Geiser, who served as director of the school between 1881 and 1887 and again between 1891 and 1895—turned to the courts to gain customs duties so as to obtain federal financial support for most of the needed expansion (Guggenbühl 1955, pp. 101–104; Oechsli 1905, pp. 344–346; and, on the customs duties, Crew 1895, diary entries entitled "Zurich—20, 21 + 22nd July 1895"). Yet Weber needed more than judicial help to obtain his institute.

In addition to the support of Kappeler and Geiser (with whom Einstein would eventually register for seven courses, including those on differential geometry, to which he later referred as "true masterpieces of the pedagogic art"), the decisive figure who advanced Weber's hopes for a new institute was Werner Siemens (Weiss 1912, p. 47; Einstein 1987, p. 44, citing Einstein 1956, pp. 10–11). Siemens was a physicist and electrotechnologist who believed that important technological advances were based on fundamental scientific advances. He was, moreover, the founder and leader of Siemens & Halske, one of the world's largest and most technologically progressive electrical firms, as well as one of the central figures in establishing the Reichsanstalt, and a close friend and relation of Helmholtz's

(Cahan 1989, pp. 24–58). When Siemens visited Zurich in 1886, a meeting was arranged among himself, Weber, Kappeler, and Geiser to discuss the proposed new physics institute. Weber envisioned a large physics institute that stressed electrotechnology, and Siemens's support for Weber's vision proved decisive in winning the backing of Kappeler and Geiser. In 1886 the federal government approved the initial funding of over 500,000 francs for the new institute (Weiss 1912, p. 47; Guggenbühl 1955, p. 104).

Weber spent much of the next four years (1887–90) constantly overseeing the institute's construction and equipping, about whose progress he kept Helmholtz regularly apprized. In July 1887, for example, he sent him "a small picture of my future institute" while the latter was still under construction (Oechsli 1905, p. 346; Weber to Helmholtz, 10 July 1887, HHN). He again wrote Helmholtz in June 1888, this time to report that he would soon be coming to Berlin, where he would visit various workshops, and that he hoped to speak with Helmholtz during his visit to the capital (Weber to Helmholtz, 17 June 1888, HHN). In October 1888, at the time of Kappeler's death, he wrote to Helmholtz of his psychological burdens and responsibilities in constructing the institute:

> I am right in the middle of constructing the physical institute, insofar as all the interior furnishings are still to be arranged. Moreover, as of next year the ongoing management of [the institute's] work must be seen to. Hence the loss of Mr. Kappeler is an extraordinarily heavy blow to me. Mr. Kappeler had devoted himself with full enthusiasm to my cause and enjoyed in this country such full respect that everything was more easily achieved with his help. (Weber to Helmholtz, 21 October 1888, HHN)

While in the 1870s Weber's mentor, Helmholtz, had guided him scientifically, now in the late 1880s Helmholtz served as a professional advisor and colleague to whom Weber could unburden himself.

The finished institute was big and expensive. Purchase of the land for and construction of the physics building cost about 1.2 million francs, while laboratory facilities and instruments cost an additional 500,000 francs (Lasius 1905, p. 336; Guggenbühl 1955, p. 104). Weber built an institute that sought, when it opened its doors in 1890, to meet

> all the demands regarding working facilities that scientific research in and teaching and application of physics require, and, indeed, in the most complete and appropriate way fulfill, including providing numerous, extended working spaces and the investigative means thus needed for introducing the entire course of the School's individual sections into practical, physical work. (Schweizerischer Bundesrat 1889, pp. 54–55)

Above all, Weber's institute was especially designed to train electrical engineers or applied physicists. That was what the Swiss wanted their tax money spent on.

Figure 2. Zurich Polytechnic Physics Institute, Physics Building (Lasius 1905, p. 336)

To be specific, the new four-story institute building contained no less than forty-two laboratories (fourteen on each of three floors), three lecture halls, six rooms for apparatus, a library, offices for the professors and lecturers, and numerous rooms to hold equipment and special apparatus. Moreover, it contained

> a complex of 4 subterranean rooms lying 8 m. below the terrace in front of the building, and intended for the execution of all those exact scientific works whose success is tied to maintaining a completely uniform temperature. Since these subterranean rooms are also built without any trace of metallic iron, they thus also maintain, in addition to almost absolute temperature constancy, the still further, and extremely valuable, advantage for finer electric and magnetic measurements of being free from all the disturbances that the presence of metallic iron carries with it. (Schweizerischer Bundesrat 1889, p. 55)

In the basement and on the ground floor Weber reserved twenty-six of the forty-two laboratories for scientific research: two for optics, two for general radiation studies, eight for mechanics and heat, ten for electromagnetic studies, and four for the professor's personal use. For these laboratories alone he received 191,000 francs for apparatus and instruments (Schweizerischer Bundesrat 1889, p. 55). On the second floor he alloted 310 square meters or six rooms—one for each of the main areas of physics—for introductory student physics laboratories, which allowed thirty to thirty-six students to work at a time. For these laboratories he received an additional 35,000 francs for apparatus and instruments (Schweizerischer Bundesrat 1889, p. 55). For his third set of laboratories, those for electrotechnology, he allotted (mostly on the second floor) 900 square meters broken up into thirteen laboratories, each of which was devoted to a particular aspect of electrotechnology and each of which was provided with appropriate electrotechnological apparatus and measuring instrumentation. (Students were expected to work their way through each of these thirteen laboratories.) Finally, Weber received a further 180,000 francs for machinery, apparatus, and instrumentation for the electrotechnical laboratories (Schweizerischer Bundesrat 1889, p. 55). In 1895, the Northwestern University physicist, Henry Crew, who was then touring Europe's physics institutes, recorded in his diary that the Zurich physics institute had "the most complete instrumental outfit I have ever seen but also the largest building I have ever seen for a physical laboratory." In the new institute he saw

> tier on tier of storage cells, dozens and dozens of the most expensive tangent & high resistance galvanometers. Reading telescopes of the largest & most expensive form by dozens, 2 or 3 in each room: ammeters, voltmeter[s], [– – – –], dynamos without end. (Crew 1895, entry entitled "Zurich—20, 21 + 22nd July 1895," cited in McCormmach 1976, p. xvi)

The situation had scarcely changed circa 1900, when the institute had an estimated budget for equipment alone of some 9,300 marks and a total budget of 17,000 marks, which meant that, among 155 of the world's physics institutes, it had the 23rd highest budget (Forman, Heilbron, and Weart, 1975, pp. 59–63).

Weber naturally needed help in operating an institute of this enormous size. When Kappeler had first brought him to Zurich, Weber found Albert Mousson, the Polytechnic's professor of experimental physics and physical geography since 1855, awaiting him. When Mousson retired in 1878, he was replaced by Heinrich Schneebeli. But Schneebeli died in May 1890, as the new physics institute opened its doors, and Weber needed a replacement (Guggenbühl 1955, pp. 234, 237). His favor fell upon Johannes Pernet (1845–1902), who, like Mousson and Schneebeli, was a native Swiss.

Unlike his predecessors, however, Pernet had led a peripatetic, financially insecure existence abroad before finally ending up at the Zurich Polytechnic. He had begun his studies of physics, mathematics, and meteorology in 1864 in Berne, and then in 1868 spent a year with Franz

Figure 3. Zurich Polytechnic Physics Institute, Physics Building (Lasius 1905, p. 317)

Neumann in Königsberg, where great emphasis was placed on precision physics, more specifically, on understanding and using precision-measuring instruments. After three years as an assistant at the St. Petersburg Central Observatory, Pernet returned for three further semesters of work under Neumann (1872–73). Then in 1874 he switched again, this time to become Oskar Emil Meyer's assistant at Breslau, where he was promoted in physics in 1875 with a dissertation on thermometry. A year later (1876) he habilitated there in physics and meteorology. His acquired expertise in thermometry, which owed a great deal to Neumann and his students, led to his call to Berlin in 1876 to work in Wilhelm Foerster's Normal-Aichungs-kommission. His expertise and Foerster's connections led to his appointment in 1877 as a *savant étranger* at the new Bureau international des poids et mesures at Bretueil near Paris, which he helped establish and where he remained through 1885 (Thiesen 1902, pp. 128–132; Wiebe 1902, p. 61; Olesko 1991, pp. 266–267, 322, 390, 392, 456, 471).

Figure 4. Zurich Polytechnic Physics Institute, Physics Building, Laboratory (Lasius 1905, p. 344)

Failing to receive the directorship of the Bureau international—Pernet reportedly managed to anger one of its directors—he returned to Berlin where, despite again finding work in Foerster's Aichungskommission, he and his family (a wife and three children) lived in extremely difficult financial circumstances (Thiesen 1902, p. 132). In 1886, Pernet again habilitated, this time at Berlin, with a dissertation that so impressed Helmholtz that he named him one of the first members of his spanking new

Reichsanstalt. Indeed, the new Reichsanstalt actually set up its temporary quarters in Pernet's private laboratory. Pernet, one of the world's foremost precision-measuring physicists, above all in the field of thermometry, now became one of Helmholtz's minions in seeing to the construction and equipping of the new institute (particularly the Physical Section) between 1887 and 1890, that is, in the very years in which Weber was constructing his new physics institute at the Polytechnic in Zurich (Thiesen 1902, p. 132; Wiebe 1902, pp. 61–62; Cahan 1989, p. 93). Moreover, Pernet befriended Helmholtz's young, sickly son Robert, himself a physicist who became associated with the Reichsanstalt before his death in 1889 (Pernet to Anna von Helmholtz, 29 December 1890, HHN). Pernet, in other words, had become a family acquaintance as well as one of Helmholtz's most trusted subordinates. With his financial future assured by his permanent position at Helmholtz's Reichsanstalt, Pernet continued his thermometric (and barometric) investigations. To all appearances, he had at last found definitive employment and security under Helmholtz's patronage. But then in May 1890 came Schneebeli's death, and Pernet received the call to return to Switzerland as Schneebeli's replacement as professor of experimental physics at the Polytechnic. No doubt Helmholtz had played an important intermediary role here.

He soon was asked to become an arbiter, too. In a letter written to Helmholtz's wife, Anna, before his very first semester at the Polytechnic had ended, Pernet alluded to problems with his colleague there, indeed to unresolvable differences. As he told Anna:

> I must press for the separation of institutes [NB] and must propose the equipping of my division. In case the officials, who basically agree with me, cannot take vigorous steps, then I shall have to try to return to Germany. (Pernet to Anna von Helmholtz, 29 December 1890, HHN)

Having already left one job, at the Bureau international, to return to Berlin, this was clearly no idle threat. Indeed, his frustrations and complaints about the situation at Weber's institute were so great, that a day later (30 December 1890) he spelled them out *ad nauseam* in a long, six-page letter to the master himself. Before doing so, however, he thanked Helmholtz, as he had already thanked Anna, for sending a picture of Helmholtz, which he placed on his desk. The picture, he wrote,

> should encourage me in these presently, and quite extraordinarily, difficult situations to lead the inexorable struggle both vigorously and also *in a factual* and *calm* manner.

Now to the causes of Pernet's nausea. He had a heavy teaching load, he claimed, and had been sick during the first semester. "How physics will develop here," he stated, "must be decided in the course of the next month." He continued:

> *In an attested declaration* my colleague unfortunately maintains the view: "At the Polytechnic there is only one physicist, and that's me" and he apparently intends to maintain this view in both theory and practice. Now, I did not come to Zurich in order to eliminate myself from the list of physicists. I have sought to meet my colleague's wishes with respect to training in the more advanced parts for regular students under his leadership, indeed, I have even made myself available in case he should require the help of a trained observer in his work. On the other hand, I indeed have an equal [*coordinirte*] position as teacher and researcher, and more lecture and laboratory students than he does. Thus, I have never recognized the situation sketched in the enclosed and, with regard to any future reorganization, I have reserved further recommendations and maintained my freedom while colleague W. has always stubbornly held the view that the conduct of scientific work *is solely his affair* [*lediglich seine Sache sei*]. Had I sufficient space and instrumentational means, then my colleague's view would leave me completely indifferent, since it is in no way shared by the officials.

Pernet then presented Helmholtz with a one-page account of the differences in his and Weber's assets. Weber, he reported, had forty-two laboratory rooms assigned to him in the institute, and Pernet was allotted only ten of these, seven of which were for general student laboratory practice work, one of them his own private laboratory, and two of which were his assistants' laboratories. Moreover, Weber reportedly had instrumentation under his control worth 323,158 francs, while Pernet had a mere 25,579 francs worth, all of which was intended for student exercises. Pernet continued:

> Colleague W. wanted to allow me and my assistants the eventual use of the so-called scientific laboratories only after an application made several months beforehand, and did not want to allow *at all* independent work executed by students under my leadership; even the use of scientific instruments for this purpose *was not allowed even in my rooms*! He has now at least relented so that students who have first completed the so-called *Diplomarbeit* with him and who later want to be involved in special work that falls in my field may do so if the work is designated as being expressly executed in my *private* laboratory.
>
> With a situation like this I have to demand, in the interest of my fully equal [*coordinirten*] division, either the use of scientific laboratories *following prior application and responsibility on my part* for the students who want to

> do scientific work under me, or, with the greatest possible consideration of all my colleague's special wishes and allowing the lion's share of the credits [to him], room distribution at least in the ratio of 2:1, along with a *definite agreement* on a corresponding equipping of my division. But my colleague seeks, by contrast, to oppose all this and to thwart thereby the planned reorganization of metrological work in weights and measures. As concerns the granting of extra credits, I certainly do not believe that possible because of the infringements, the building, etc., which may be charged to my colleague. Hence, I also do not believe in [any] essential remedy, all the less so since the same colleague W. completely fights against that. If a solution is not found that permits me a corresponding scientific utilization of my knowledge corresponding to the very pronounced needs here, then the support which has been shown to me and my general stimulation would not be enough to keep me from returning to Germany, and eventually applying [for a position there].

His horrendous institutional situation notwithstanding, Pernet added that he had nonetheless learned much in Zurich because he was now required to teach all fields of physics and because he had to familiarize himself with much electrical equipment. He noted, too, that for his lectures on experimental physics he recommended Heinrich Kayser's textbook, "which in many points follows your lectures" (Pernet to Helmholtz, 30 December 1890, HNN; for the photograph, see Pernet to Anna von Helmholtz, 29 December 1890, HHN and Pernet 1895, pp. 3–4).

All of this occurred in Pernet's very first semester at Zurich (1890–91), and it might be supposed that, after he had a chance to settle in (and down), and after he and Weber had become better accustomed to one another, the situation improved. Yet this was apparently never to be. It seems that Pernet was interested in a new position in February 1894, for Helmholtz requested a copy of Pernet's *vita* and list of publications. He was perhaps thinking about nominating Pernet to succeed Hertz, who had died the previous month, at Bonn (Pernet to Helmholtz, 14 February 1894, HHN). Crew's visit to the institute in July 1895 shows that the relationship between Pernet and Weber had not improved. Crew recorded in his diary:

> The best of feeling does not exist between these two men so Prof. [Karl] Ritter [of the Polytechnic] tells me—this is also evident from their conversation. The result is that they have duplicated each other's outfit at frightfull [*sic*] cost. . . . Pernet & Weber have each 3 assistants—[these] two mechanics & several apprentices—besides numerous services.

This, in addition to the institute's expensive plant, was why Crew spoke of a "frightful waste of money." Crew recorded in 1895 that "Pernet says they have 300 students . . . since these 300 come to the laboratory from engineering, chemical and agricultural depts.—while [only] 3 are really interested in physics" (Crew 1895, entry entitled "Zurich—20, 21 + 22nd July 1895").

Pernet had come to the Polytechnic in the expectation of finally achieving professional independence, of heading or at least co-heading his own institute where he could teach and do research as he saw fit. He was bitterly disappointed. Thanks to his heavy teaching load, he accomplished only a very limited amount of research during his twelve years (1890–1902) at the Polytechnic: a few studies on thermometry, rotating comparator, X-rays, and the specific heat of water (Thiesen 1902, p. 133; Wiebe 1902, p. 62). Nonetheless, at the heart of all these studies, just as at the heart of Weber's, was precision measurement and instrumentation. Moreover, Pernet had returned to Switzerland in the hope of undertaking a magnetic survey of the country and of establishing a miniature Reichsanstalt, a Swiss metrological center aimed at testing instruments just as Helmholtz's new Reichsanstalt and Foerster's older Normal-Aichungskommission in Berlin and the older Bureau international in Breteuil were doing (Pernet 1891a, esp. p. 6; Pernet 1891b; Pernet 1894a; Pernet 1894b; Wiebe 1902, p. 62; Thiesen 1902, p. 133). Here, too, he was disappointed, and his program for pursuing measuring physics remained confined within the small space allotted him by Weber. Moreover, with the death of Helmholtz in September 1894, Pernet lost his patron and probably any chance of leaving Zurich and escaping Weber. Helmholtz's death so moved Pernet that he wrote a thirty-six-page obituary essay. Pernet delivered his encomium before the Naturforschende Gesellschaft in Zurich in December 1894. He wrote:

> May these lines prove that in our country too, the country which Helmholtz so loved, the country of Euler, Bernoulli, and Albrecht von Haller, his work is honored, and that the breach opened by his death is likewise painfully felt [here as in Germany itself]. (Pernet 1895, p. 4)

Pernet concluded his peroration by referring to Helmholtz as "the greatest natural researcher of our century," and, his death notwithstanding, "to the everlasting efficacy of his ideas and accomplishments, and to the love, honor, and gratitude of all those who had the good fortune of knowing him personally and who will always mourn for [the loss of] one of the noblest of men" (Pernet 1895, 36). While the master's body had died, his spirit lived on to inspire Pernet, who spent all too much of the last seven years of his life, until his own death in 1902, in ill-health (Thiesen 1902, p. 133; Wiebe 1902, p. 62).

4. The Young Einstein's Physics Education

As this essay has sought to show, during the decade prior to Einstein's arrival at the Zurich Polytechnic, Weber had established a magnificent physics institute, one dedicated above all to teaching and training in precision measurement and in applied physics, in particular electrotechnology. Weber offered his young wards virtually unmatched physical facilities and a well-trained, experienced staff.

In October 1896, Einstein immatriculated at the Zurich Polytechnic, enrolling in the section for *Fachlehrer* of mathematics and natural science, which, during his four years there (1896–1900) registered between forty-five and fifty-six students per semester (Oechsli 1905, p. 367; Guggenbühl 1955, p. 256). The section had two subdivisions, with Einstein's (VIA) then headed by Hurwitz; in 1896, it had only twenty-three registered students (Einstein 1987, p. 43). The section's enrollment was thus small enough that students belonging to it could easily come to know the staff. Two years later (October 1898), Einstein passed his qualifying examination (*Übergangs-Diplom-Prüfung*), receiving a "5 ½" in "Physics" ("6" being the highest grade possible, "1" the lowest).

Einstein could henceforth concentrate on physics during his remaining two years at the Polytechnic. He now came into regular contact with the institute's two professors of physics, Weber and Pernet. This institute, then, constituted Einstein's formal, immediate educational environment between 1898 and 1900, and the remainder of this essay seeks to explore the ways in which the institute in general, as well as the personalities of Weber and Pernet and the writings of Helmholtz and others, influenced the young Einstein's understanding and practice of physics. In particular, it explores, first, how Weber's teaching of and general attitude towards the nature of physics may have influenced Einstein (both positively and negatively); second, how Weber introduced or helped introduce Einstein to several specific physics fields, issues, and results that proved of noteworthy importance to the young Einstein's research efforts; and third, how Weber affected the young Einstein on his path to becoming a professional physicist.

By all accounts Weber was an excellent teacher. Even the hostile Pernet conceded as much. He wrote to Helmholtz in 1890 that, due to illness, Weber "has had to sit out almost the entire semester, to the great sorrow of his listeners, for he gives excellent lectures" (Pernet to Helmholtz, 30 December 1890, HHN). He inspired students, at least forty-three of whom earned their doctorates under him (Weiss 1912, p. 51). He did so in good part by teaching by example, presenting elementary laws of physics and concrete experimental results, as Einstein's lecture notes abundantly reveal (Einstein 1987, pp. 63–210). Pierre Weiss, for his part, asserted that Weber's "simultaneously elegant and precise lectures exerted on all who [had the opportunity] to hear them, an unforgettable impression. For one who heard Weber lecture, physics was an affair of revelation." Weiss also noted that, in his physics instruction, Weber showed little interest in the "general point of view." Yet he made physics come alive for his listeners: "It was no more a dead science," Weiss said, "it was called into life by him" (Weiss 1912, p. 50). He certainly brought it to life for Einstein during his second year at the Polytechnic. In February 1898, Einstein wrote to Mileva Marić, his fellow student and future wife: "Weber lectured on heat (temperature, quantity of heat, heat movement, the dynamic theory of gases) with great mastery. One lecture after another of his pleases me" (Einstein 1987, p. 212). Again, over a half-century later Einstein wrote a correspondent: "I valued Weber as a very gifted teacher, one who gave the essential of things and economically limited the diversity" (quoted in Einstein 1987, p. 60). As Table 1 shows, Einstein took no fewer than four year-long (i.e., two-semester) courses with Weber and seven one-semester courses. Thus, Einstein registered for no fewer than fifteen courses with Weber! And his grades for those courses in which grades were given were high: three "5"s ("Physics," "Scientific Work in the Physical Laboratory," and "Scientific Work in the Physical Laboratories"), a "5 ½" ("Physics") and two "6"s ("Electrotechnology Laboratory" and "Scientific Work in the Physical Laboratories") (Table 1; cf. Einstein 1987, pp. 46–48, 60, 366–369).

Five of Einstein's courses with Weber were laboratory courses. Weber supervised his laboratories very carefully, so carefully that he occasionally even rebuked students who failed to put in enough time in the laboratory. He was reportedly the first one in the laboratory in the morning and the last to leave at night; no one could avoid him (Weiss 1912, p. 51). No doubt, then, he could greet Einstein as he entered and left the laboratory. For although Einstein was often enough bored with his lecture classes and more than a little prone to cutting them, he loved to work in Weber's superbly-equipped laboratories, which stood at Einstein's disposal while Weber preached the importance of measurement in physics and practiced measuring physics. Einstein wrote: "I worked . . . in Professor H.F. Weber's physical laboratory with fervor and passion" (Einstein 1956, pp. 10–11).

Table 1. Einstein's Courses and Grades with Weber and Pernet at the Zurich Polytechnic from the Winter Semester (WS) 1897–98 to the Summer Semester (SS) 1900 (Einstein 1987, pp. 46–48)

Course	**Instructor**	**Semester**	**Grade**
Physics	Weber	WS, 1897–98	5½
Physics	Weber	SS, 1898	5
Principles, Apparatus, and Measuring Methods of Electrotechnology	Weber	WS, 1898–99	–
Electrical Oscillations	Weber	WS, 1898–99	–
Electrotechnology Laboratory	Weber	WS, 1898–99	6
Introduction to Physical Practicum	Pernet	WS, 1898–99	–
Physical Practicum for Beginners	Pernet	WS, 1898–99	1
Principles, Apparatus, and Measuring Methods of Electrotechnology	Weber	SS, 1899	–
Electrotechnology Laboratory	Weber	SS, 1899	–
Scientific Work in the Physical Laboratory	Weber	SS, 1899	5
Introduction to Electromechanics	Weber	SS, 1899	–
Alternating Currents	Weber	SS, 1899	–
Alternating Current Systems and Alternating Current Motors	Weber	WS, 1899–1900	–
System of Absolute Electrical Measurements	Weber	WS, 1899–1900	–
Scientific Work in the Physical Laboratories	Weber	WS, 1899–1900	6
Scientific Work in the Physical Laboratories	Weber	SS, 1900	5
Introduction to the Theory of Alternating Currents	Weber	SS, 1900	–

And again: "I worked . . . the most time in the physical laboratory, fascinated by the direct contact with experience" (Einstein 1949, p. 14).

During the summer semester of 1899, while working in one of Weber's laboratory courses, Einstein's enthusiasm, or carelessness, led to his seriously wounding his hand (Einstein 1987, pp. 218, 222). Even after graduating from the Polytechnic, Einstein intended to use Weber's laboratory to conduct experimental work in thermoelectricity. In late August or early September 1900, he wrote Marić of his plans: "At all costs we must try to stay in good stead with Weber, since his laboratory definitely is the best and the best equipped" (Einstein 1987, p. 258).

If Weber often had the pleasure of greeting Einstein in the laboratory, Pernet rarely did. In the winter semester 1898–99, Einstein took his only two courses with Pernet: "Introduction to Physical Practicum," for which he received no grade, and "Physical Practicum for Beginners," for which he received a "1"—the only "1," indeed the only low grade, that he received at the Polytechnic (Table 1; Einstein 1987, pp. 46–50). The teacher's conference for Einstein's section recorded: "Due to neglect of the physical practicum, Einstein receives, on the written instruction of Herr Pernet, a reprimand from the [Polytechnic's] Direction"; and the Polytechnic's registrar recorded in Einstein's official transcript: "Reprimand by the Director due to lack of diligence [*Unfleiss*] in the physical practicum" (Einstein 1987, pp. 47 [quotes], 60, 368; see also Kollros 1956, p. 21, and Seelig 1960, pp. 48–49). Einstein clearly had a poor relationship with Pernet (see Seelig 1960, p. 65, and the reference cited in Einstein 1987, p. 60, n. 2). His failure to attend Pernet's laboratory regularly, his low mark and official reprimand from Pernet, and the fact that he received uniformly high marks in Weber's courses suggest the possibility that the animosity and rivalry between Pernet and Weber may have spilled over onto Einstein. Did Einstein think of himself as Weber's, and not Pernet's, student? Did Pernet perceive Einstein as Weber's student and project his hostility toward Weber onto Einstein?

Against Weber's many pedagogical strengths stood one glaring weakness, failing to teach Maxwell's electromagnetic theory. His failure to transmit that theory to his students—most of whom were future physics teachers, electrotechnologists, and professional physicists—shows that by the late 1890s, when Maxwell's theory had become widely accepted, Weber had become seriously deficient in teaching, if not understanding, contemporary electromagnetic theory and its applications. Einstein later recalled: "The most fascinating subject at the time that I was a student was Maxwell's theory" (Einstein 1949, p. 33). Einstein's fellow student Louis Kollros put the point judiciously and pertinently:

> His [Weber's] lectures on classical physics were lively [*lebendig*], but we waited in vain for a presentation of Maxwell's theory. We knew that it confirmed the identity of transmission of electricity and light and that Hertz's investigations on electric waves had confirmed the theory. Einstein above all was disappointed. (Kollros 1956, p. 22)

Another close Einstein friend and Zurich Polytechnic student, Michele Besso, later wrote to him of a young student of Besso's who was "naive, as was the old H.F. Weber with regard to Maxwell" (Speziali 1972, pp. 315–316). And Adolf Fisch, who studied at the Aarau Kantonschule and the Polytechnic when Einstein did, later wrote that Weber's

> lectures were outstanding. But as a typical representative of classical physics he simply ignored everything that came after Helmholtz. After concluding studies, one knew the past of physics, but not its present and future. We were thus left to study the newer literature privately. (As quoted in Seelig 1960, p. 47)

Einstein, for one, did just that: He capitalized on Weber's failure by turning to self-study, namely, to the writings of Helmholtz, Maxwell, Kirchhoff, Hertz, Boltzmann, and to the recent textbooks of August Föppl or Paul Drude (Kollros 1956, p. 22; Einstein 1987, p. 223).

For Einstein, furthermore, Weber had a perhaps still greater shortcoming than his failure to teach Maxwell: He did not teach the foundations of physics, and he did not teach theoretical or mathematical physics. To learn these aspects of physics, as well as to learn Maxwell's theory, Einstein had only his instinctive autodidacticism to turn to. Apart from Weber's occasional remarks on theory in his lectures, Einstein received no training in theoretical physics at the Polytechnic. By and large, virtually no courses in theoretical physics as such were taught during his final two years at the Polytechnic and, to the extent they were, Einstein was unable to take them (Kollros 1956, p. 21; Einstein 1987, pp. 43, 264, 265n). (Still, it is noteworthy that in his *Diplom* examination [July 1900] Einstein scored "10" [out of "12"] in "Theoretical Physics" as well as "10" in "Practical Physics") (Einstein 1987, p. 247). His courses in mathematical physics—in particular those in mechanics with Albin Herzog, and potential theory and analytical mechanics with Minkowski—were the nearest that he came to formal instruction in theoretical physics (Einstein 1987, pp. 46–49, 364–365, 367, 369). After a lecture on capillarity by Minkowski during Einstein's final semester at the Polytechnic, he told Kollros: "That's the first lecture on mathematical physics that we've heard at the Poly!" (Kollros 1956, p. 21). Yet he apparently made something of it, for capillarity is the the subject of Einstein's first published physics paper, submitted to the *Annalen der Physik* in December 1900, less than five months after receiving his *Diplom* (Kollros 1956, p. 23; Einstein 1901; Einstein 1989, pp. 3–6). Moreover, Einstein's early (pre-1907) attitude towards the role of mathematics in physics may also have owed at least something to that of Weber, who, as already noted, had limited mathematical ability. As Einstein later wrote: "it was not clear to me as a student that the approach to a more profound knowledge of the basic principles of physics is tied up with the most intricate mathematical methods. This dawned upon me only gradually after years of independent scientific work" (Einstein 1949, pp. 15–17).

If Weber failed to teach theoretical and mathematical physics, he and his institute at least helped further sensitize Einstein to the importance of measurement for testing theory and for finding the best fit between theory and empirical reality. At the very least, Weber drove home the importance of carefully using precision instrumentation in physics, and he provided Einstein with the necessary, indeed outstanding, instrumentation of the Polytechnic's sumptuous laboratories.

No doubt other influences also led Einstein to appreciate the importance of precision instrumentation and measurement. Just as the Einstein family's electrotechnical firm doubtless sensitized Albert to the importance of electrotechnology before he entered Weber's institute (Pyenson 1982), so, too, it might be argued, Einstein initially was sensitized to handling precision instruments starting at age six, when he began eight years worth of violin lessons. In 1896, a music examiner for Einstein's class at the Aargau Kantonsschule reported that in violin the class "as a whole performed very satisfactorily, both technically and in matters of intonation," and that "one pupil, by the name of Einstein, even performed brilliantly, as shown by his profoundly understood rendition of an Adagio from a Beethoven sonata" (Einstein 1987, p. 21). In music, as in physics, Einstein was never satisfied with mere technique; instead, he sought in both rhapsodic pleasure from an underlying harmony (Einstein 1987, pp. lviii, 21, 219, 370–371). His long hours in Weber's laboratory, like his music lessons, sensitized Einstein to the delicacy of instruments in general, to the beauty and harmony that they could create, when properly used, and to the importance of caring for one's instrument. This sensitivity to precision instruments and their use in assessing theory against physical reality is one of the principal features of the young Einstein's writings (see Hentschel 1992 for Einstein's later attitude toward experiments and instruments). In April 1901, he wrote to Marić: "I am again presently studying Boltzmann's gas theory. Everything is very beautiful, except too little value is laid on the comparison with reality. I believe, however, that in O.E. Mayer there is to be found enough empirical material for our investigation" (Einstein 1987, p. 294). This remark epitomizes his scientific style; and his reference to Oskar Emil Meyer, Pernet's *Doktorvater* from Breslau and the author of *Kinetische Theorie der Gase*, suggests how Einstein's own contemporary work on the applications of his theory of molecular forces to transport phenomena in gases belonged to a rich scientific tradition (Einstein 1987, pp. 265, 292).

While Weber and, to a far lesser extent, even Pernet belonged to that tradition, there were of course others to whose writings the young Einstein turned for self-instruction, who influenced his understanding of physics. Above all, he could, and did, benefit from the work and general scientific outlook of Helmholtz. When Einstein was only sixteen years old, he made an independent and fundamental study of Jules Violle's *Lehrbuch der Physik*, underlining and annotating the text (Einstein 1987, pp. lxiv, 6). His study of Violle's textbook already owed something to Helmholtz, for Violle's *Lehrbuch* had been translated into German in 1892–93 by Ernst Gumlich, Ludwig Holborn, Wilhelm Jaeger, Damian Kreichgauer, and Stephan Lindeck, all of whom were then employed at the Reichsanstalt, where they served Helmholtz (Violle 1892–93). While at the Polytechnic Einstein often enough did not attend his lecture classes (probably those in mathematics), but instead "studied at home, and with holy zeal, the masters of theoretical physics" (Einstein 1956, p. 10). Among them, he related, were "the works of Kirchhoff, Helmholtz, Hertz, etc." (Einstein 1949, p. 14). Indeed, even when he lacked access to Zurich's libraries, he was not prevented from reading Helmholtz. During a visit to his parents in Milan in September 1899, he planned to visit the city library to take out books by Helmholtz (as well as by Boltzmann and Mach) and promised Marić that they would later review everything together when he returned to Zurich (Einstein 1987, p. 230).

As this essay has argued, Helmholtz had played key roles in the careers of Weber and Pernet, from providing them with a laboratory, to offering them teaching experience in his institute, to conducting research on specific heats and precision measurement in general, and to helping Weber gain his post at the Polytechnic and Pernet his at the Reichsanstalt and doubtless at the Polytechnic as well. Both men looked to Helmholtz as a patron and intellectual leader; he was a sort of *éminence grise* of the Polytechnic physics institute, and it seems reasonable to assume that they emphasized his name and work to their students.

Einstein, moreover, had an additional source from which he may have learned to admire Helmholtz. Einstein regularly visited the home of Alfred

Stern, the professor of history at the Polytechnic; Stern became a close friend and adviser to Einstein (Einstein 1987, pp. 216, 246, 296–297, 298; Seelig 1960, pp. 185–187). Stern, who was the son of the Göttingen mathematician Moritz Stern, may well have told Einstein how, while a student of law and history at the University of Heidelberg, he attended science courses given by Kirchhoff and Helmholtz (Einstein 1987, p. 386). In particular, at Heidelberg around 1865 he attended Helmholtz's lectures on "The General Results of the Natural Sciences," which were meant for students of all disciplines. Stern wrote that

> the hours in which Helmholtz, in informal but also stimulating and intellectually exciting lecture, treated the most various of subjects—the sensations of tone, the construction of the human eye, spectral analysis, the skulls of Neanderthals—made an indelible impression on me for my entire life. (Stern 1932, p. 5)

Stern, who was also a friend of Aaron Bernstein, the author of *Naturwissenschaftliche Volksbücher*, a work that Einstein studied as a youth "with breathless attention," may well have passed on that sense of Helmholtz to Einstein (Stern 1932, p. 11; Einstein 1949, p. 15; Einstein 1987, p. lxii).

Helmholtz's direct and indirect effects on Einstein can also be seen in several of the important physical fields and problems that had preoccupied Helmholtz during his career and later preoccupied the young Einstein: understanding the laws of thermodynamics and electrodynamics, clarifying the mechanical foundations of thermodynamics, and reformulating, through a reformation of mechanics, the foundations of physics as a whole. Einstein wrote to Marić in August 1899:

> I've returned the Helmholtz volume and am presently once again studying Hertz's *Propagation of Electric Force* very carefully. The occasion for this was that I did not understand Helmholtz's treatise on the principle of least action in electrodynamics. I am becoming more and more convinced that the electrodynamics of moving bodies, as it is currently presented [i.e., in a paper by Hertz], does not correspond to reality and instead can be presented more simply. The introduction of the name "ether" into electrical theories has led to the idea of a medium of whose movement one may be able to speak, yet in my opinion without being able to link a physical sense to this expression. (Einstein 1987, p. 226)

Six years later, Einstein would, of course, publicly present a full-scale, rigorous rendition of this conviction—providing a simpler theory that corresponded better to reality and without invoking an ether. Yet this letter to Marić was only Einstein's first linkage of Helmholtz's name with problems occasioned by electrodynamics. Writing a month later (late September 1899) from his parents' home in Milan, he told her to prepare to read Helmholtz with him upon his return to Zurich: "let's begin with Helmholtz's *Electromagnetic Theory of Light*," he wrote, "which I have still not read 1) due to anxiety and 2) because I didn't have it" (Einstein 1987, p. 235). Some two weeks later (October 1899), while still visiting his parents in Milan, he again instructed Marić: "In the meantime definitely get Helmholtz's electromagnetic theory of light! I already really have a hunger for it" (Einstein 1987, p. 238). In the meantime, he had satisfied his hunger by studying Helmholtz's writings on atmospheric motions. In early August 1899 he wrote to Marić: "I have . . . also studied something very beautiful in Helmholtz on atmospheric movements," noting that he wanted to discuss this entire subject further when Marić returned to Zurich (Einstein 1987, p. 220; Helmholtz 1888). He added: "When I read Helmholtz the first time, I couldn't understand at all that you weren't sitting by me, and I'm not doing any better now. I find working together very good and healthy and, besides, less dry" (Einstein 1987, p. 221).

At the heart of Helmholtz's philosophy of science lay the concept of causality, a priori organizing principles within a neo-Kantian framework, an emphasis on measurement in science, a quest to establish the foundations of physics and to do so on the basis of so-called principle theories of physics, and, finally, the mechanical foundations of science. With the obvious exception of this last, these concepts and concerns also lay at the heart of Einstein's philosophy of science. Even Einstein's language seems similar to Helmholtz's. As Martin Klein, in speaking of Einstein's 1902 paper on the kinetic theory of heat equilibrium and on the second law of thermodynamics, has written: "Einstein's language, when he refers to the mechanical representation of a physical system or to the mechanical world picture, is the language of the previous generation, of Hermann von Helmholtz and Heinrich Hertz, of J.J. Thomson and J. Willard Gibbs" (Klein 1982, p. 40). Hertz's language, it may be noted, is in good measure Helmholtz's. The similarities in language, concepts, choice of research fields, and general philosophy of science between Helmholtz and Einstein should leave it as no surprise to hear Einstein report to Marić in 1899: "More and more, I wonder at the original, free thinker Helmh[oltz]" (Einstein 1987, p. 220).

Weber also influenced Einstein, introducing him or guiding him along to several specific physics fields, issues, or results that proved important to his first research efforts. Weber was not, of course, by any means the only physicist to influence Einstein; this essay has already noted several of the physicists who constituted the scientific tradition that helped shape the young Einstein's physics research program and his thought more broadly. Nonetheless, Weber's role in that tradition is tangible and merits brief notice.

First, Einstein's notes from his two-semester, general physics course (1897–98) with Weber show an especial concern with thermodynamics, electricity, and magnetism, while at the same time revealing a surprising amount of attention to precision-measuring details for an introductory, general course. As shown by the notes of Emil Teucher, who was Einstein's fellow student in Weber's course and who took more extensive notes than Einstein did, Weber's lectures were also concerned with thermometry and barometry (Einstein 1987, p. 62). Moreover, Einstein's own notes show that Weber also lectured on other topics that involved measurement and instrumentation, including the spherometer, calorimeter, electrometer, galvanometer, and Wheatstone bridge (Einstein 1987, pp. 75, 88–92, 156–158, 162–164, 172, 186–187). The young Einstein, as is well known, was particularly concerned with thermodynamics and its mechanical foundations, and the bulk of Einstein's notes concern thermodynamics (Einstein 1987, pp. 63–147). Furthermore, Weber's lectures seem to be Einstein's first detailed, quantitative introduction to the issue of intermolecular forces, a topic that he pondered over between 1900 and 1902 and that resulted in his first publications (Einstein 1987, pp. 62, 123, 130, 261, 264–266, 285, 290–292, 295, 303, 320, 324; McCormmach 1970, p. 43; Einstein 1901, 1902). Moreover, Einstein's choice of heat conduction as the topic of his *Diplomarbeit* reflected Weber's own (and old) interests more than those of Einstein, who showed little enthusiasm for this traditional type of study.

Einstein's introduction to the problem of blackbody radiation may also have been due to Weber. In his course during the summer semester of 1899, "Principles, Apparatus, and Measuring Methods of Electrotechnology," Weber introduced his students to the very advanced topic of blackbody radiation and its applicability to understanding problems concerning electric lighting (Einstein 1987, p. 197n). In 1904, the young Einstein's analysis of the general molecular theory of heat explicitly cited Wien's displacement law, which itself derived from Weber's semi-empirical radiation law (Einstein 1904, pp. 360–362).

Weber's own research results also found their way into the young Einstein's first independent research. Weber's work on the specific heats of carbon, boron, and silicon, showing their anomalous temperature dependency in contrast to the predictions of the Dulong-Petit rule, remained a puzzle until Einstein, in 1907, used the quantum hypothesis to explain that anomalous behavior and pointed explicitly and specifically to data first found by Weber in the early 1870s (Einstein 1907, pp. 185, 190; Pais 1982, pp. 394–397).

Third and finally, Weber deeply affected the young Einstein as he sought to become a professional physicist. Einstein's professional relationship with Weber began quite positively, yet ended quite negatively. As a mere sixteen-year-old pupil, Einstein went to Zurich to take the entrance examination for the Polytechnic. He was then two years under the normal age for immatriculation; held Württembergian, not Swiss, citizenship; and had left his Munich Gymnasium without obtaining his leaving certificate. Although he failed the examination, Weber found his performance on the scientific part of the examination so strong that he encouraged Einstein to attend his lectures; he spotted the young Einstein's talent and gave him encouragement (Einstein 1956, pp. 9–10). Einstein chose instead to recommence his secondary-school studies, this time at the Aargau Kantonsschule, which, thanks to its rector, the physicist August Tuchschmid, one of Weber's former assistants at the Polytechnic, had an outstanding laboratory (Einstein 1987, pp. 10–11; Einstein 1956, p. 9).

During most of his three years of study under Weber's supervision (1897–1900), Einstein apparently had a good relationship with him. As already discussed, his grades under Weber were uniformly high and he enjoyed working in Weber's laboratory. When did the tensions between the two begin, and where did they come from? Did they stem from Einstein's disappointment at Weber's failure to teach Maxwell's theory and other parts of contemporary physics? Or, on a more personal level, did they stem from Einstein's practice of referring to Weber as "Herr Weber" instead of "Herr Professor Weber" (Seelig 1960, p. 48)? Given Weber's authoritarian treatment of his colleague Pernet, it is easy enough to imagine what little indulgence he might have had for a seemingly disrespectful student who was a foreigner and a Jew to boot. Be that as it may, Einstein did do his *Diplomarbeit* under Weber, though with an *experimental* topic on heat conduction in which he had little interest (Einstein 1987, pp. 244, 327). He graduated from the Polytechnic in July 1900, receiving a diploma qualifying him to teach secondary-school mathematics and physics (Einstein 1987, pp. 44, 50). Yet when Einstein and his friends Jakob Ehrat, Marcel Grossmann, and Kollros received their diplomas, Einstein alone among them failed to receive, as he had so hoped, the offer of an assistantship. Rather than take Einstein, Weber took two mechanical engineers. Kollros believed that Einstein was "doubtless seen as too independent" by Weber (Kollros 1956, p. 22). The memory of this bitter experience with Weber never left Einstein: In 1934 he recalled that his professors "didn't like me because of my independence, [and] avoided me when they needed assistants"; and as late as 1952, two years before his death, he still remembered that "at the last moment an assistantship was not given to me" (as quoted in Einstein 1987, p. 44).

Graduation did not quite bring an end to Einstein's professional relationship with Weber. For one, Einstein planned to write a doctoral dissertation (*Doktorarbeit*) with Weber, and he apparently spent the winter of 1900–01 doing so. He originally intended to work on thermoelectricity, a topic that Weber would have welcomed; instead, however, he switched to molecular forces, a topic that could scarcely have pleased Weber. Whether this change in topic is cause or effect of his hostile relationship with Weber is unclear. What is clear is that the planned dissertation under Weber came to naught (Einstein 1987, pp. 61, 258, 259, 269–270, 272, 290).

For another, having been passed over by Weber for an assistantship at the Polytechnic, Einstein sought to strike out on his own. He tried to get one with, among others, Hurwitz at the Polytechnic, Otto Wiener in Breslau, Wilhelm Ostwald in Leipzig, and Eduard Riecke in Göttingen. To Hurwitz, he declared that "no opportunity had been offered for seminar exercises in theoretical and practical physics.... In my student years I was mainly occupied with analytical mechanics and theoretical physics" (Einstein 1987, p. 264). He did not get the position (Einstein 1987, pp. 262, 263–264, 269n, 253). To Wiener, he announced that he had learned his physics "through attendance at lectures, study of the classics, as well as through work in the physical laboratory" (Einstein 1987, p. 277). He did not get that post. To Ostwald, the physical chemist, Einstein portrayed himself as "a mathematical physicist who is intimately familiar [*vertraut*] *with absolute measurements*" (Einstein 1987, p. 278, emphasis added). His desire to work with Ostwald was not merely that of landing a job: prior to writing Ostwald, Einstein told Marić that he was "completely delighted about the successes" in physical chemistry during the past thirty years (Einstein 1987, p. 267), a field that Helmholtz had helped initiate (Kragh 1993). Einstein told Ostwald that the latter's *Lehrbuch der allgemeinen Chemie* had greatly excited him (Einstein 1987, p. 278). Yet Ostwald was apparently unimpressed: he answered neither Einstein's letter nor his follow-up letter, nor that of Einstein's father, Hermann, on Albert's behalf (Einstein 1987, pp. 279, 284, 285, 287, 289–290; see also Körber 1964). Finally, Einstein's bid to work under Riecke had, in principle, no hope of success since Riecke's job announcement called for someone with a doctorate (Einstein 1987, p. 279, n. 2). Yet Riecke was an old friend of Einstein's close friend and mentor Stern (Stern 1932, p. 17), and Einstein may perhaps have used Stern's name or hoped for Stern's influence with

Riecke. Be that as it may, he believed that Weber was responsible for his lack of success here (if not in all the other cases as well). He wrote Marić:

> That's a malicious story [*eine böse Geschichte*] with Riecke; I think the position is almost certainly lost. I scarcely believe that Weber will let so beautiful an opportunity go by without doing something [against me]. Following your advice, my love, I have written to Weber so that he will at least know that what he does he cannot do behind my back. I wrote him that I know that my appointment still depends on his recommendation alone. (Einstein 1987, p. 279)

Einstein's failure to get the assistantship with Riecke turned Weber into his nemesis. Einstein wrote Marić in March 1901:

> Riecke's refusal hasn't surprised me. And I'm also completely convinced that Weber is guilty here. The excuse is simply too improbable; and he hasn't even mentioned anything at all about the second position.
>
> I am convinced that under these circumstances it no longer makes sense to write further to professors, since, should things get far enough along, it is certain that they would all enquire with Weber, and he would again give a negative report. I'll turn to my old teachers in Aarau and Munich; mainly, however, I'll try to get an assistantship in Italy. In the first place, here [i.e., in Italy] the main difficulty—namely, the anti-Semitism—does not occur, which in the German-speaking lands I find as unpleasant as it is troublesome; in the second place, here I have some protection. (Einstein 1987, pp. 281–282)

Whether or not Weber was an anti-Semite, Einstein was surely correct to believe that being Jewish hurt his chances and that without Weber's help he had no chance for an academic position in the German-speaking world. Two weeks after he wrote Marić, he told Grossmann: "I might have long ago found such a [position] if Weber had not played falsely with me" (Einstein 1987, p. 290). Still, for the qualities desired in an assistant, Weber may well have had objective grounds for passing over Einstein. As Einstein himself noted, he was anything but a model student. He lacked, he later said, "a quick comprehension" and "willingness to concentrate his energies on all that is lectured on; orderliness, so as to provide a written inventory of all that was said in the lectures and then to work it up conscientiously." Instead, he was "a sort of vagabond and eccentric" (Einstein 1956, pp. 10–11). Those were not the qualities Weber and others needed in their assistants.

Even if Weber did undermine Einstein's attempts to get an assistantship, training in his institute nonetheless proved to be of much professional value to Einstein. For the training Einstein received in electrotechnology and applied physics there proved crucial to him as he sought his first jobs. In his application to work at the Technikum Burgdorf, where he was employed from May to July 1901, he wrote of his studies at the Polytechnic: "In addition to the usual mathematical and physical disciplines, I also heard there technical fields, like the strength of materials with Prof. Hertzog [*sic*] and electrotechnology with Prof. Weber" (Einstein 1987, p. 307). And in his application for a position at the Patent Office, where he worked from 1902 to 1909, he wrote:

> I acquired my specialist knowledge in the field of physics and electrotechnology at the school for specialists in the mathematical-physical direction at the Eidgenössisches Polytechnikum in Zurich. . . . There I received at the end of my studies, on the basis of an experimental study in the field of physics and a submitted test, the federal diploma for specialists [*das eidgenössische Diplom für Fachlehrer*]. (Einstein 1987, p. 327)

Finally, if Weber's institute were nothing else for Einstein, it was at least a center for meeting other students with whom he developed deep social and intellectual ties. Above all, attending Weber's classes brought Einstein into contact and friendship with the conscientious Grossmann, who did more than anyone to help Einstein pass his examinations, obtain his first permanent job (in June 1902, at the Patent Office), acquire the mathematical tools he needed to formulate the general theory of relativity, and who, finally, became the key figure in arranging for Einstein's eventual appointment as professor of theoretical physics at the ETH in 1912 (Einstein 1956, pp. 11–12, 15–16; Kollros 1956, pp. 23, 27). Therein lay some irony. Twelve years earlier Einstein could not even beat out a mechanical engineer for an assistantship with Weber. Now he stood, along with Weiss, who was Weber's successor, as one of the two leaders of the Zurich physics institute. Indeed, he stood alone at the pinnacle of theoretical physics and perhaps at the height of his intellectual powers, and though Weber was now dead and buried, Einstein's heart had so hardened against him that he could write to Heinrich Zangger: "Weber's death is good for the ETH" (quoted in Pais 1982, p. 45).

5. Conclusion

The historiographic spectrum for analyzing the young Einstein must be a broad one. On the one end, there can be no doubt that Einstein's reading of philosophers like Hume, Kant, and Mach were essential components in his intellectual development that gave him general intellectual stimulation and helped him focus on the philosophical and conceptual foundations of science. On the other end, there can be little doubt that the rise of the electrotechnical industry in general and the business activities of Einstein's father and uncle in particular (Pyenson 1982) were also essential components in the young Einstein's development, just as the general business activity of Switzerland helped provide the institutional wherewithal for creating the Zurich Polytechnic's physics institute. Yet it seems extremely difficult to determine precisely what such readings and business activities contributed to his development.

By contrast, focus on the institutional environment, in particular on the Zurich Polytechnic physics institute, where the young Einstein received his physics education, provides a natural focus that occupies the middle ground of the historiographical spectrum. As this essay has argued, Weber's institute constituted a large share of the concrete social, material, and intellectual environment out of which the young Einstein emerged. Here the young Einstein learned some of the elements of physics; here in the person of Weber he found a gifted teacher of experimental physics, an experienced if erstwhile researcher, and an entrepreneurial if authoritarian institute leader; here he met the disgruntled metrologist Pernet; here he came to appreciate and admire the writings of Helmholtz and other leading scientists, and to understand the scientific tradition of which he was soon to be an important part; here he received expert laboratory instruction or gained through his own wit invaluable laboratory experience, and here he handled the finest physical equipment, apparatus, and instrumentation of the day that helped sensitize him to the importance of precision measurement; and finally, here he came to know and love stimulating fellow students—Marić, Kollros, and Grossmann, to name but a few. Weber and his institute gave Einstein much. To be sure, he and it failed to give Einstein what he craved most—guidance in theoretical physics and a vision of the grand unity of the field. But then, perhaps not even a Helmholtz could have given the young Albert Einstein all that he craved.

Acknowledgments. An early version of this essay was presented to the conference on "Einstein: The Early Years," 4–6 October 1990, at the Boston University Conference Center at Osgood Hill, North Andover, Massachusetts, and I should like to thank the conference's organizers, Don Howard and John Stachel, and participants for their comments, as well as Klaus Hentschel and Kathryn M. Olesko. I thank the Berlin-Brandenburgische Akademie der Wissenschaften Archiv for permission to quote from letters in Hermann von Helmholtz's Nachlaß (HHN), Nr. 496; the Northwestern University Library, Evanston, for permission to quote from the diary of Henry Crew; and the Bibliothek of the Eidgenössische Technische Hochschule, Zurich, for permission to reprint the photograph of H.F. Weber.

NOTES

[1] Einstein had only three months worth of physics education (under Joseph Ducrue) at the Luitpold gymnasium in Munich (Einstein 1987, p. 353).

[2] Weiss stressed the influence of Abbe and, to a lesser extent, the philosopher Kuno Fischer, but he noted only in passing that of Helmholtz (Weiss 1912, pp. 44–45).

[3] Pierre Weiss 1912, pp. 52–53, provided a bibliography of Weber's writings, which should be supplemented by the items listed in Poggendorff 1898 (vol. 3) and 1904 (vol. 4), s.v. "Weber, Heinrich Friedrich."

[4] It should be noted, however, that in July 1887 he wrote to Helmholtz that he had a manuscript in hand on "electrical oscillations in open circuits of small capacity and limited self-induction . . . (theoretical and experimental)." Hertz's recent publications had reminded him of this, and Weber hoped that Helmholtz might consider this manuscript for publication at the Berlin Akademie (Weber to Helmholtz, 10 July 1887, HHN). Weber apparently never published this manuscript.

REFERENCES

Auerbach, Felix (1922). *Ernst Abbe. Sein Leben, sein Wirken, seine Persönlichkeit*, 2nd ed. Leipzig: Akademische Verlagsgesellschaft.

Cahan, David (1985). "The Institutional Revolution in German Physics, 1865–1914." *Historical Studies in the Physical Sciences* 15: 1–65.

— (1989). *An Institute for an Empire: The Physikalisch-Technische Reichsanstalt 1871–1918*. Cambridge, New York, and New Rochelle: Cambridge University Press.

— ed. (1993). *Hermann von Helmholtz and the Foundations of Nineteenth-Century Science*. Berkeley, Los Angeles, and London: University of California Press.

Crew, Henry (1895). Diary: "Notes of Travel. Europe 1895." Henry Crew Papers. Northwestern University Archives. "Diaries— European Trips 1895, 1921, 1927." Box 6, Folder 3.

Drude, Paul (1894). *Physik des Aethers auf elektromagnetischer Grundlage*. Stuttgart: Ferdinand Enke.

Einstein, Albert (1901). "Folgerungen aus den Capillaritätserscheinungen." *Annalen der Physik* 4: 513–523.

— (1902). "Über die thermodynamische Theorie der Potentialdifferenz zwischen Metallen und vollständig dissociirten Lösungen ihrer Salze und über eine elektrische Methode zur Erforschung der Molecularkräfte." *Annalen der Physik* 8: 798–814.

— (1904). "Zur allgemeinen molekularen Theorie der Wärme." *Annalen der Physik* 14: 354–362.

— (1907). "Die Plancksche Theorie der Strahlung und die Theorie der spezifischen Wärme." *Annalen der Physik* 22: 180–190.

— (1949). "Autobiographical Notes." In *Albert Einstein: Philosopher-Scientist.* Paul Arthur Schilpp, ed. Evanston, Illinois: The Library of Living Philosophers, pp. 2–94.

— (1956). "Autobiographische Skizze." In Seelig 1956, pp. 9–17.

— (1987). *The Collected Papers of Albert Einstein.* Vol. 1, *The Early Years, 1879–1902.* John Stachel, et al., eds. Princeton: Princeton University Press.

— (1989). *The Collected Papers of Albert Einstein.* Vol. 2, *The Swiss Years: Writings, 1900–1909.* John Stachel, et al., eds. Princeton: Princeton University Press.

Föppl, August (1894). *Einführung in die Maxwell'sche Theorie der Elektricität.* Leipzig: B.G. Teubner.

Forman, Paul, Heilbron, John L., and Weart, Spencer (1975). "Physics *circa* 1900: Personnel, Funding, and Productivity of the Academic Establishments." *Historical Studies in the Physical Sciences* 5.

Frey-Wyßling, A., and Häusermann, Elsi, comp. (1955). *Geschichte der Abteilung für Naturwissenschaften an der Eidgenössischen Technischen Hochschulen in Zürich 1855–1955.* N.p.: N.p., n.d.; copy in ETH Bibliothek, Zurich.

Guggenbühl, Gottfried (1955). *Geschichte der Eidgenössischen Technischen Hochschule in Zürich.* [Zurich]: Buchverlag der Neuen Zürcher Zeitung.

Helmholtz, Hermann von (1880). "Über Bewegungsströme am polarisierten Platina." *Annalen der Physik und Chemie* 11: 737–759.

— (1888). "Über atmosphärische Bewegungen." *Königlich Preussische Akademie der Wissenschaften* (Berlin). *Sitzungsberichte*: 647–663.

— (1889). "Über atmosphärische Bewegungen. (Zweite Mitteilung). Zur Theorie von Wind und Wellen." *Königlich Preussische Akademie der Wissenschaften* (Berlin). *Sitzungsberichte*: 761–780.

Hentschel, Klaus (1992). "Einstein's Attitude towards Experiments: Testing Relativity Theory 1907–1927." *Studies in History and Philosophy of Science* 23: 593–624.

Kangro, Hans (1970). *Vorgeschichte des Planckschen Strahlungsgesetzes. Messungen und Theorien der Spektralen Energieverteilung bis zur Begründung der Quantenhypothese.* Wiesbaden: Franz Steiner.

Klein, Martin J. (1982). "Fluctuations and Statistical Physics in Einstein's Early Work." In *Albert Einstein: Historical and Cultural Perspectives. The Centennial Symposium in Jerusalem.* Gerald Holton and Yehuda Elkana, eds. Princeton: Princeton University Press, pp. 39–58.

Körber, Hans-Günther (1964). "Zur Biographie des jungen Albert Einstein. Mit zwei unbekannten Briefen Einsteins an Wilhelm Ostwald vom Frühjahr 1901." *Forschungen und Fortschritte* 38: 74–78.

Kollros, Louis (1956). "Erinnerungen." In Seelig 1956, pp. 17–31.

Kragh, Helge (1993). "Between Physics and Chemistry: Helmholtz's Route to a Theory of Chemical Thermodynamics." In Cahan 1993, pp. 403–431.

Kremer, Richard L. (1993). "Innovation through Synthesis: Helmholtz and Color Research." In Cahan 1993, pp. 205–258.

Lasius, G. (1905). "Die Gebäude der Eidgenössischen polytechnischen Schule." In *Festschrift zur Feier des fünfzigjährigen Bestehens des Eidg. Polytechnikums.* Part 2, *Die bauliche Entwicklung Zürichs in Einzeldarstellungen von Mitgliedern des Zürcher Ingenieur- und Architektenvereins.* Zurich: Polygraphisches Institut and Zürcher & Furrer, pp. 321–346.

McCormmach, Russell (1970). "Einstein, Lorentz, and the Electron Theory." *Historical Studies in the Physical Sciences* 2: 41–87.

— (1976). "Editor's Foreward." *Historical Studies in the Physical Sciences* 7: xi–xxxv.

Oechsli, Wilhelm (1905). *Festschrift zur Feier des fünfzigjährigen Bestehens des Eidg. Polytechnikums.* Part 1, *Geschichte der Gründung des Eidgenössischen Polytechnikums mit einer Übersicht seiner Entwickelung 1855–1905 von Wilhelm Oechsli.* Frauenfeld: Huber.

Olesko, Kathryn M. (1991). *Physics as a Calling: Discipline and Practice in the Königsberg Seminar for Physics.* Ithaca, N.Y. and London: Cornell University Press.

Pais, Abraham (1982). *'Subtle is the Lord . . .': The Science and the Life of Albert Einstein.* Oxford, New York, and Toronto: Oxford University Press.

Pernet, Johannes (1891a). "Über die physikalisch-technische Reichsanstalt zu Charlottenburg und die daselbst ausgeführten elektrischen Arbeiten." *Schweizerische Bauzeitung* 18: 1–6.

— (1891b). Pernet to Heinrich von Bötticher, Staatssekretär im Reichsamt des Innern, Berlin, Zürich-Hottingen, 6 August 1891, "Die Errichtung einer Physikalisch-technische Reichsanstalt," 13144/3, Staatsbibliothek Preußischer Kulturbesitz (formerly Deutsches ZentralArchiv) Potsdam, Reichsamt des Innern.

— (1894a). "Über den Einfluss physikalischer Präzisionsmessungen auf die Förderung der Technik und des Mass- und Gewichtswesens." *Schweizerische Bauzeitung* 24: 110–114.

— (1894b). "Über die Förderung der Schweiz. Technik durch die eidg. Eichstätte und das eidgen. physikalische Institut." *Schweizerische Bauzeitung* 24: 116–118, 121–123.

— (1895). "Hermann von Helmholtz 31. August 1821 bis 8. September 1894. Ein Nachruf." *Neujahrsblatt der Naturforschenden Gesellschaft in Zürich* 97: 1–36.

Poggendorff, J.C. (1898). *J.C. Poggendorff's Biographisch-Literarisches Handwörterbuch zur Geschichte der exacten Wissenschaften*. Leipzig: J.A. Barth.

— (1904). *J.C. Poggendorff's Biographisch-Literarisches Handwörterbuch zur Geschichte der exacten Wissenschaften*. Leipzig: J.A. Barth.

Pyenson, Lewis (1982). "Audacious Enterprise: The Einsteins and Electrotechnology in Late Nineteenth-Century Munich." *Historical Studies in the Physical Sciences* 12: 373–392.

Schweizerischer Bundesrat (1889). *Die Eidgenössische Polytechnische Schule in Zürich*. Ed., im Auftrage des Schweizerischen Bundesrathes bei Anlass der Weltausstellung in Paris 1889. Zurich: Zürcher & Furrer.

Seelig, Carl (1960). *Albert Einstein. Leben und Werk eines Genies unserer Zeit*. Zurich: Europa Verlag.

— ed. (1956). *Helle Zeit—Dunkle Zeit. In Memoriam Albert Einstein*. Zurich, Stuttgart, and Vienna: Europa Verlag.

Speziali, Pierre, ed. (1972). *Albert Einstein—Michele Besso. Correspondance, 1903–1955*. Paris: Hermann.

Stern, Alfred (1932). *Wissenschaftliche Selbstbiographie*. Zurich and Leipzig: Gebr. Leemann.

Thiesen, Max (1902). "Nachruf für Johannes Pernet." *Deutsche Physikalische Gesellschaft. Verhandlungen* 4: 128–135.

Violle, Jules (1892–93). *Lehrbuch der Physik*. 2 vols. E. Gumlich et al., eds. Berlin: Julius Springer.

Weber, Heinrich Friedrich (1872). "Die spezifische Wärme des Kohlenstoffs." *Annalen der Physik und Chemie* 147: 311–319.

— (1875). "Die spezifischen Wärmen der Elemente Kohlenstoff, Bor und Sicilium." *Annalen der Physik und Chemie* 154: 367–423, 553–582.

— (1877). "Absolute elektromagnetische und calorische Messungen." *Naturforschende Gesellschaft in Zürich. Vierteljahrsschrift* 22: 273–322.

— (1878). "Die Inductionsvorgänge im Telephon." *Naturforschende Gesellschaft in Zürich. Vierteljahrsschrift* 23: 265–272.

— (1879). "Untersuchungen über die Wärmeleitung in Flüssigkeiten." *Naturforschende Gesellschaft in Zürich. Vierteljahrsschrift* 24: 252–98. Reprinted in *Annalen der Physik und Chemie* 10 (1880): 103–129.

— (1881). "Die Beziehung zwischen dem Wärmeleitungsvermögen und dem elektrischen Leitungsvermögen der Metalle." *Königlich Preussische Akademie der Wissenschaften* (Berlin). *Monatsberichte*: 457–478.

— (1884). *Der absolute Wert der Siemensschen Quecksilbereinheit*. Zurich: Zürcher und Furrer.

— (1885). "Das Wärmeleitungsvermögen der tropfbaren Flüssigkeiten." *Königlich Preussische Akademie der Wissenschaften* (Berlin). *Sitzungsberichte*: 809–815.

— (1886a). "Kritische Bemerkungen zu den neuen Entdeckungen von Hughes über die Selbstinduktion in metallischen Leitern." *Zentralblatt für Elektrotechnik*: 162–163.

— (1886b). "Die Selbstinduktion bifilar gewickelter Drahtspiralen." *Königlich Preussische Akademie der Wissenschaften* (Berlin). *Sitzungsberichte*: 511–524.

— (1887a). "Zur Theorie der Wheatstone'schen Brücke." *Annalen der Physik und Chemie* 30: 638–655.

— (1887b). "Die Entwicklung der Lichtemission glühender fester Körper." *Königlich Preussische Akademie der Wissenschaften* (Berlin). *Sitzungsberichte*: 491–504. Reprinted in *Annalen der Physik und Chemie* 32: 256–270.

— (1888). "Untersuchung über die Strahlung fester Körper." *Königlich Preussische Akademie der Wissenschaften* (Berlin). *Sitzungsberichte*: 933–957.

— (1897). "Berücksichtigung der Formen der Wechselstrom-Spannungen und Wechselstrom-Intensitäten bei den Messungen von Capazitäten und Induktionscoefficienten mittelst Wechselstrom." *Annalen der Physik und Chemie* 63: 366–375.

Weiss, Pierre (1912). "Prof. Dr. Heinrich Friedr. Weber. 1843–1912." *Schweizerische Naturforschende Gesellschaft. Verhandlungen* 95: 44–53.

Wiebe, Hermann (1902). "Professor Dr. Johannes Pernet." *Deutsche Mechaniker-Zeitung* 7: 61–63. Reprinted in *Zeitschrift für Instrumentenkunde* 22.

Kant's Impact on Einstein's Thought

Mara Beller

"Aber das Genie wird eben nie schülerhaft abhängig"

— August Stadler

1. Introduction

In his "Autobiographical Notes," Einstein summarized his life's effort as an unremitting struggle to establish a new, unified foundation for physics (Einstein 1946). The epistemological analysis of scientific concepts and structures formed an integral part of this struggle. As Einstein himself said:

> The reciprocal relationship of epistemology and science is of noteworthy kind. They are dependent upon each other. Epistemology without contact with science becomes an empty scheme. Science without epistemology is—insofar as it is thinkable at all—primitive and muddled. (Einstein 1949, pp. 683–684)

No wonder historians and philosophers of science have invested a considerable amount of effort to uncover the epistemological roots of Einstein's science and philosophy. The roots of Einstein's philosophical approach have been traced to many thinkers–Mach, Helmholtz, Poincaré, Hume, Duhem and Spinoza.[1] Kant's name is not among them.

The most widely accepted analysis of Einstein's philosophical makeup in the historical literature, at least until recently, was that by Gerald Holton (1968), who described Einstein's pilgrimage from his early positivism à la Mach to the rationalist position of his later years in the spirit of Spinoza. In the analysis of Einstein's philosophy one is hardly surprised to find strands of many, sometimes contradictory epistemological opinions. Einstein himself, one might recall, described a working scientist as an "epistemological opportunist" (Einstein 1949, p. 684).

The central claim of this paper is that there is more coherence and consistency in Einstein's philosophical outlook throughout the years than might initially appear. The main thesis will be that the Kantian framework

Einstein Studies, vol. 8: Einstein: The Formative Years, pp. 83–106.

enables us to get a fresh insight into Einstein's philosophy of science and scientific style, and to realize in what way different, seemingly incompatible aspects of Einstein's philosophy hang together. In particular, Einstein's commitment to determinism, his view of the relation between theory and experiment, and what seems an inconsistent blend of realism and conventionalism, will appear in a new light.

To be sure, there are significant differences between Einstein and Kant—Einstein's rejection of the Kantian *synthetic a priori* is perhaps the most prominent among those differences. Admittedly, there is also a change of emphasis from Einstein's early empiricist spirit to his position in later years. This later position is, I suggest, more adequately characterized by the term "neo-Kantian," rather than as a form of rationalism along the lines of Descartes and Spinoza.

While acknowledging the differences between Einstein and Kant, this paper aims to uncover important similarities between the philosophies of Kant and Einstein. The relevance of the Kantian framework for the understanding of Einstein's thought is based less on the specifics of the Kantian system, and more on Kant's method and general approach.

But what is our evidence for Einstein's familiarity with Kantian philosophy? In a sense, one almost does not need such evidence. The prevalence of Kantian philosophy among the educated classes in Germany in the late nineteenth century, and the extent to which it was taught in German universities, was overwhelming.[2] One way or another, the enlightened German citizen could not but absorb the basic ideas of the Kantian system.

In Einstein's case, we can make a much stronger claim. Einstein considered Kant not only a great philosopher, but also an outstanding spiritual leader of mankind.[3] He read and reread Kant's writings many times throughout his life. Einstein first read Kant at the age of thirteen; his older friend, the poor Polish medical student Max Talmey, recommended to Einstein, among other books, the *Critique of Pure Reason*. Another childhood friend from Einstein's Swiss high school reminisced that Einstein read Kant again at the age of sixteen (Howard 1994, p. 49). From his ETH transcript, we know that Einstein was probably exposed to Kantian teachings again in the year 1897, when he enrolled in lectures on Kant's philosophy by August Stadler, the distinguished neo-Kantian of the Marburg school (Einstein 1987, pp. 45–50, 362–369). And there is evidence that Einstein took an active part in discussions on Kant in 1911–12, during his time at the Charles University in Prague (Howard 1994, p. 49). In 1918 Einstein was immersed in Kant again, as he wrote to Max Born: "I am reading Kant's *Prolegomena* here, among other things, and am beginning to comprehend the enormous suggestive power that emanated from the fellow and still does. . . . Anyway it is very nice to read, even if it is not as good as his predecessor Hume's work" (Born 1971, pp. 25–26). Of course, Einstein's exposure to Kant was also indirect, through those classics of nineteenth century physics that Einstein read during his student days at the ETH, among them Helmholtz and Poincaré.

While we have extensive evidence of Einstein's early encounter with Kant, we do not know which parts of the Kantian system were prominent in Einstein's understanding of Kant. Stadler's lectures on Kant therefore constitute an important and reliable source for the young Einstein's exposure to the main themes of the Kantian framework.

A few words about Stadler: The Stadler Nachlass is stored in twenty-three folders in the central library of Zürich (Zentralbibliothek). The text looks very strange, with missing vowels, and is probably written in South-German dialect. The notes for the lectures on Kant that Stadler delivered at the ETH are located in folder number 12. After Stadler's death, a professor named Platter apparently worked through Stadler's notes, and published a large portion of them in book form. Included among these is a book with the title *Kant. Akademische Vorlesungen*, whose table of contents is identical, chapter-by-chapter, to the list of contents in folder number 12.[4] It is then a fair assumption that one can rely on the published version as a faithful guide to those Kantian ideas that Einstein encountered in Stadler's lectures.

The most pervasive theme in Kant's approach, Stadler emphasizes, is the realization that any construction of a speculative system must be preceded by a thorough analysis of the concepts with which the system operates. The analysis of concepts and a careful examination of the domains and limits of their applicability should definitely precede any constructive efforts. The historical way in which Stadler presented the development of Kantian thought reinforces this point again and again. As Stadler emphasized, whenever Kant encountered a problem, his approach was to subject the categories that he employed to the most stringent examination, and to reduce them to as small a number as possible. Even the most common, the most obvious of our concepts, should not escape this critical scalpel: "One of the philosopher's most important tasks is to ask of everything that is supposed to be self-evident whether or not it really is self-evident. . . . He must destroy the illusion [*Wahn*] of self-evidence" (Stadler 1912, p. 150).

This warning to beware of premature speculative efforts, of the danger of building on "shaky foundations," of the necessity of a thorough analysis of concepts, found direct echo in the work of the young Einstein. But because this approach could have been learned by Einstein elsewhere, notably from Mach's books, I prefer to concentrate on two other central themes in Stadler's lectures, which are uniquely Kantian. The first theme is the dual theory of intuition and judgment, which acknowledges that both empirical and conceptual components are essential for knowledge.

The second theme, which is of utmost importance for understanding Einstein's thought, is Kant's concept of the systematic unity of Nature as a regulative idea. This idea gives meaning to the concept of scientific lawfulness, truth, objectivity, causality and God's order. The Kantian concept of systematic unity is a methodological maxim as well, which defines the meaning of, and motivation for, scientific research.

2. Dual Theory of Intuition and Judgment

Einstein's philosophical approach throughout most of his intellectual life was that of distancing himself from what he called "sterile positivism," stating emphatically that the basic laws and concepts of a scientific system cannot be deduced from experience. Yet neither was he a rationalist in the spirit of Descartes and Spinoza, as is sometimes claimed: He always acknowledged the empirical aspect of knowledge—"all knowledge of reality starts in experience and ends in it"—emphasizing repeatedly that "pure logical thinking cannot yield us any knowledge of the empirical world" (Einstein 1933, p. 271). What Einstein called "the eternal antithesis between the two *inseparable* components of our knowledge—the empirical and the rational" (Einstein 1933, p. 271), is indeed the essence of the Kantian dual theory of intuition and judgment. Kant, challenging both "naive empiricism" and "dogmatic" rationalism, is the originator of the philosophical position that acknowledges that both empirical and conceptual components are essential for knowledge and are present in every single representation.

This aspect of Kantian philosophy constitutes perhaps the central theme of Stadler's lectures on Kant. Valid theoretical knowledge is knowledge only of that which is empirically perceived, teaches Stadler, yet a purely empiricist philosophy is not sufficient: In order to gain knowledge one must use pure mental constructs—forms of understanding—which are not derivable from phenomena. Whatever is given to us is given "only through sensory perception," yet "without thinking, sensory perception never comes clearly into consciousness" (Stadler 1912, pp. 188–189). No valid knowledge, Stadler emphasizes, can lead us beyond the world of phenomena. Even though our thought operates with purely mental constructs, which do not originate in experience, these constructs find no valid theoretical application beyond the sphere of experience. Their origin is not empirical, yet only empirical use can be made of them. What Stadler presents as a unique philosophical achievement of the *Critique of Pure Reason* is Kant's demonstration that neither raw sense data nor purely logical thinking are sufficient; we need purely mental constructs: "Merely logical thought . . . can order and clarify our experiences, but cannot create experience. For that we need concepts of the pure understanding" (Stadler 1912, p. 197).

Einstein's philosophical writings contain many passages that are very close in spirit to the Kantian approach to the relation between sense-data and our conceptual apparatus. According to Einstein, the method that the physicist follows involves basing one's foundations on some general principles and deducing conclusions from these principles. One cannot learn how to deduce these general principles from experience; one can only get at them by revelation ("principles . . . revealed themselves to him" [Einstein 1914, p. 221]), by intuitively arriving at certain general postulates that the scientist somehow "sniffs out" from the comprehensive complexes of empirical facts. Until the physicist finds these principles, "individual empirical fact is of no use" to him; moreover, "he cannot even do anything with isolated experimental and general laws abstracted from experience" (Einstein 1914, p. 221).

This dual theory of knowledge is eloquently expressed in Einstein's Herbert Spencer Lecture at Oxford in 1933. Einstein's position is opposed to that of the rationalist: "Propositions arrived at by purely logical means are completely empty as regards reality" (Einstein 1933, p. 271). Nevertheless, the fundamental concepts and laws of the theoretical system are not derivable from experience—they are the work of reason, free inventions of the human mind; yet the justification of these principles lies solely in the empirical consequences of the theory as a whole.

Einstein's emphasis on the necessity of prior, organizing principles is not strictly Kantian, of course: They are not a priori, but are free inventions, legitimized only by the empirical success of the theoretical system. Still, this Kantian duality of the "rational" and the "empirical" is fundamental for Einstein's philosophy of science.

The priority of some conceptual notions for the very possibility of experience is also acknowledged by Einstein in his debate with the neo-Kantians in the 1920s. Thus, Einstein wrote to Ernst Cassirer on 5 June 1920: "I acknowledge that one must approach the experiences with some sort of conceptual functions, in order for science to be possible; but I do not believe that we are placed under any constraint in the choice of these functions *by virtue of the nature of our intellect*" (quoted in Howard 1994,

p. 54). Einstein summarized his views with explicit acknowledgement to Kant in his "Reply to Criticisms":

> The theoretical attitude here advocated is distinct from that of Kant only by the fact that we do not conceive of the categories as unalterable. . . . They appear to be a priori only insofar as thinking without the positing of the categories and of concepts in general would be as impossible as breathing in a vacuum. (Einstein 1949, p. 674)

According to Einstein, as science advances, its logical basis becomes more unified, while the number of elementary concepts and relations among them that constitute such a basis gradually decreases. So the "story goes on until we have arrived at a system of the greatest conceivable unity, and of the greatest poverty of concepts of the logical foundations, which is still compatible with the observations made by our senses." The price for the increasing unity and logical simplicity of the theory is the growing distance between the basic theoretical principles and the "primary" concepts, that is, concepts connected with sense experiences (Einstein 1936, p. 294). In this state of affairs, mathematics plays an increasingly greater role in the search for the unified basis of physics. In his Spencer Lecture, Einstein emphasized the crucial role of mathematics in a scientist's search for a true physical theory:

> Our experience hitherto justifies us in believing that nature is the realization of the simplest conceivable mathematical laws. I am convinced that we can discover by means of purely mathematical constructions the concepts and the laws connecting them with each other, which furnish the key to the understanding of natural phenomena. . . . In a curious sense, therefore, I hold it true that pure thought can grasp reality, as the ancients dreamed. (Einstein 1933, p. 274)

It is these words that are taken as evidence for Einstein's Spinozistic rationalism in his later years (Holton 1968, p. 252). Yet from the rest of the text of this lecture, as well as from numerous discussions by Einstein elsewhere, it is clear that Einstein sees mathematics as an enormous heuristic force in the search for the correct scientific theory, as a unifying power, and not as a criterion of scientific truth. In fact, Einstein considers a rationalism in the spirit of Spinoza an aristocratic "illusion" or "prejudice":

> During philosophy's childhood it was rather generally believed that it is possible to find everything which can be known by means of mere reflection.

> It was an illusion . . . Even in Spinoza . . . this prejudice was the vitalizing force which seems still to have played the major role. (Einstein 1944, pp. 19–20)

Equally untenable is a "plebeian illusion of naive realism" (Einstein 1944, p. 20). Even though Hume (and Galileo) clearly saw that "the empirical . . . procedure alone has shown its capacity to be the source of knowledge," Hume's strongest legacy is rather his demonstration that such essential concepts as causal connection cannot be gained from material given by the senses (Einstein 1944, p. 21).[5] Hume's greatness is then in his preparation of the stage for Kant's solution:

> Then Kant took the stage with an idea which, though certainly untenable in the form in which he put it, signified a step towards the solution of Hume's dilemma: whatever in knowledge is of empirical origin is never certain (Hume). If, therefore, we have definitely assured knowledge, it must be grounded in reason itself. (Einstein 1944, p. 22)

And while dissociating himself from the Kantian "synthetic a priori," Einstein strongly endorses the general Kantian approach: "The following, however, appears to me to be correct in Kant's statement of the problem: in thinking we use, with a certain 'right,' concepts to which there is not access from the materials of sensory experience" (Einstein 1944, p. 22).

Einstein's attitude, while displaying both strong rationalist and empiricist aspects,[6] can be more fittingly characterized as a neo-Kantian one. This characterization is not a pettiness of terminology: Only by acknowledging Einstein's debt to Kant can we put into proper balance what seem to be contradictory strands in Einstein's epistemological position.

Perhaps the most illuminating Kantian notion for understanding Einstein's epistemological framework is Kant's idea of the systematic unity of nature,[7] which will put into a new, more coherent perspective the basic features of Einstein's philosophy of science, as well as of his unique scientific style.

3. Unity

We have seen that for both Kant and Einstein the only true knowledge is empirically grounded knowledge. Kant's analysis of the limits of empiricism will illuminate the central role of the concept of the "unity" or "order" of Nature in the philosophy of science of both Kant and Einstein.

In the *Critique of Pure Reason*, Kant discusses the "hypothetical employment of reason" (Kant 1787, p. 535), pointing out repeatedly that the confirmation of a hypothesis by its empirical consequences can never endow such a hypothesis with universality, or certainty: "In natural science . . . there is endless conjecture, and certainty is not to be counted upon" (Kant 1787, p. 433). The most such empirical confrontation can do is to impart to a given hypothesis a greater or lesser degree of probability. We cannot know all of the possible consequences of the hypothesis that might prove its universality—so repeated comparison of observation with deductive consequences is, as Kant emphasizes, "of little importance for science."

Experiments themselves do not represent some sort of "pure case" of empirical procedure. In experimental physics—writes Kant—even the "principles according to which we perform experiments must themselves always be derived from the knowledge of nature, and hence from theory" (as quoted in Buchdahl 1969, p. 510, n. 1). Some theoretical constructs aim not merely at increasing the probability of the given theory—a theory could not even begin without them—they are also presuppositions, according to which we perceive, construct, and create the empirical world (see Kant 1787, p. 534); compare this with Einstein's emphasis on the creative, constructive nature of scientific theorizing.

Because a given hypothesis cannot obtain the proof of its truth from "below," from repeated experimental confirmation, something else is needed. We need a criterion that will distinguish contingent and limited, unimportant, empirical generalities from laws of nature, which are endowed with universality and necessity. Paradoxically, we need a non-empirical criterion for the notion of empirical truth. So one is led to the Kantian concept of unity and to the need to systematize scientific experience.

Reason cannot rest simply on assembling a collection of particular empirical laws, claims Kant. It is partly a drive of reason, partly the very definition of what reason is, to seek a further systematization of these laws into a unified, general framework: "The hypothetical employment of reason has, therefore, as its aim the systematic unity of the knowledge of understanding, and this unity is *the criterion of the truth* of its rules" (Kant 1787, p. 535). It is not just that such systematization allows interconnections between the different, and disconnected, empirical generalities: *From the infinitely many uniformities only those that can be fitted into a unified, systematized general system can be regarded as laws, and can be regarded as having law-like necessity.*

This unity of Nature is not an ontological principle. It is meaningless, according to Kant, to ask whether Nature in fact possesses such a unity. It is rather a methodological principle, presupposed in the very idea of scientific research: Without the principle of unity we will have no reason, no science, and no sufficient criterion of empirical truth. As Kant puts it, "In order, therefore, to secure an empirical criterion we have no option save to presuppose the systematic unity of nature as objectively valid and necessary" (Kant 1787, p. 538). The epistemological importance of Kant's notion of the systematic unity of Nature is that it provides a criterion of validity for scientific hypotheses, totally different from that of the empirical idea of confirmation. This criterion is a source for the development of holistic ideas in the philosophy of science after Kant, including Kantian underpinnings in the writings of the logical positivists (see especially Friedman 1983, Friedman 1987 and Ryckman 1991). Note also that the Kantian idea of the "truth" of a statement is equivalent to its being a law-like, or necessary statement—for Kant, and later for Einstein, a true theory necessarily means a causal theory.

Much has been written about the importance of the idea of unity in Einstein's life and work, and about Einstein's psychological, or motivational attachment to this idea. Yet this principle does not simply supply the drive for doing science: It is an epistemological criterion of truth, or validity, for scientific theories repeatedly used by Einstein throughout his life. It is his commitment to this epistemological criterion that explains Einstein's numerous reactions to experimental tests that sometimes are misinterpreted as disregard of the importance of experiment, or as a demonstration of his rationalistic attitude. It is this criterion that allowed Einstein to consider certain theories true before their experimental confirmation, or other theories true in spite of their seeming experimental disconfirmation.

There are many remarks in Einstein's writings that indicate that he attached no meaning to the "truth" of scientific theory dissociated from the "unity" criterion. According to Einstein, only a theory whose object is the totality of all physical experience is a candidate for being considered a true theory: "A system has truth-content according to the certainty and completeness of its coordination-possibility to the totality of experience. A correct proposition borrows its 'truth' from the truth-content of a system to which it belongs" (Einstein 1946, p. 13).

Such a holistic approach characterizes Einstein not only in his later years. It is already present in the early stages of Einstein's career, at least since 1906–1907. It is this approach that allowed Einstein to disregard the results of Kaufmann's "crucial" experiments, which seemed to support the Abraham-Bucherer electron theory and to refute the common predictions of the Einstein and Lorentz theories (Holton 1968, p. 253; Miller 1981, p.

124). As Einstein put it then, the competing theories "have rather small probability, because their fundamental assumptions (concerning the mass of moving electrons) are not explainable in terms of theoretical systems which embrace a greater complex of phenomena" (as quoted in Holton 1968, p. 253).

The striving for unity in science is not unique to Einstein, of course. It is an ancient ideal that inspired many researchers throughout the ages. It would be far-fetched to assume that Einstein inherited this ideal only, or even primarily, from Kant. The young Einstein was a witness to heroic attempts to provide a unified foundation for physics—be it a mechanistic, electromagnetic, or energetic one (Klein 1980). The unity and interconnection of optics and electromagnetism in Maxwell's theory, which Einstein considered as the most fascinating subject of study in his student days, "was like a revelation" (Einstein 1946, p. 3).

What is unique in Einstein's case is not merely the fact that he devoted his life's effort almost exclusively to the search for unity in science. The unity of the theoretical domain served in Einstein's case as an explicit criterion of the truth of scientific theory, quite along Kantian lines.

In a series of recent articles, Don Howard attempts to refute the picture of Einstein as a simple-minded realist—a picture that Einstein's opponents in the camp of the orthodox interpretation of quantum mechanics were always eager to propagate (Howard 1990, 1993). Instead, he pictures Einstein as a conventionalist, and traces the holistic aspects of Einstein's holism to Pierre Duhem. Yet Einstein was exposed to Kant much earlier than to Duhem; moreover, Einsteinian holism seems to be totally at odds with Duhemian holism. For the aim of Duhemian holism is precisely to undermine the notion of the "truth" of a scientific theory, relegating the concept of "truth" to another, incompatible realm—that of religion (Duhem 1906); while Kantian and Einsteinian holism legitimizes and defines the very notion of scientific truth.

As I already mentioned, the concept of the unity of Nature, or order of Nature, is a methodological maxim or regulative idea. In fact, the Kantian idea of unity is not merely a methodological principle, it is simultaneously a transcendental principle, for it provides not merely methodological guidance, but the very presupposition for the development of scientific methodologies (Buchdahl 1969, p. 512). There is no "order of Nature" out there, there is no "order of Nature" without scientific research. All questions along the lines of a "correspondence theory of truth" are meaningless in this Kantian framework. The disciplined reason should therefore avoid asking questions of this sort about whether such an order exists or can be known. This systematic unity cannot be known—we have to think it (Kant). Compare this with Einstein's telling words: "It is basic for physics that one assumes a real world . . . but this we do not *know*. We take it only as a program in our scientific endeavors" (quoted in Fine 1984, p. 95).

Kant rejects not only the ontological meaning of the principle of unity, but he foreshadows—and rejects—the Machian principle of the economy of thought:

> It might be supposed that this is merely an economic contrivance whereby reason seeks to save itself all possible trouble. . . . But such a selfish purpose can be very easily distinguished from the idea. For in conformity with the idea everyone presupposes that this unity of reason accords with nature itself, and that reason . . . does not here beg but command." (Kant 1787, p. 539)

The Kantian idea of systematic unity leads further to the ideal notion of "completable" research and to the related notion of God as Divine Designer, or a "heuristic fiction." The idea of "systematic" unity is often interchanged in Kant's writings with the notion of "purposive unity," "as if" the world were constructed by a divine Designer, or an archetypal intellect, through a general plan of an interconnected and coherent system. Reason cannot comprehend this systematic unity otherwise than by assigning to the idea of this unity an object—the notion of God (Kant 1787, p. 550). The idea of a Kantian God is postulated only problematically, as a "heuristic fiction"; it is a regulative principle by which reason can extend the notion of systematic unity over the whole field of experience. Kant emphasizes repeatedly that no assertion, or even assumption, of existence, can be associated with the idea of God. It is, like the principle of systematic unity, merely a regulative idea in the pursuit of scientific investigation. This idea, to which no object in experience corresponds, can be grasped only analogically. Kant provides four equivalent analogues: methodological (as a regulative idea), teleological (as a final cause), theological (as a supreme wisdom, or supreme intelligence), and empirical. The last of these—the empirical—are the general scientific systems that serve as examples of unified science: for Kant, the theories of Maupertuis, Newton, Linnaeus; for Einstein, paradigmatic examples of classical science, such as thermodynamics and Maxwell's electrodynamics. The advantage of the idea of a supreme intelligence over the mere notion of systematic unity, despite their equivalence, is that it allows us to imagine an indefinitely extendable process of research that leads to a completed, or "total" (Kitcher 1986) science, and therefore to an ideal, ultimate corpus of systematic knowledge, which is the true and necessary, albeit empirically discovered, knowledge of the world.

There is no doubt that there is a close affinity between the Einsteinian and the Kantian God. The Einsteinian God is the God of Kant, and not of Spinoza, as often stated (sometimes by Einstein himself): It is the God of an empirically discovered interconnectedness, not of purely logical connection. Whenever Einstein uses the concept of God, he uses it as a methodological recommendation, as in "God does not play dice," directed against statistical theories being considered as fundamental, or "God is subtle but not malicious," directed against those ad hoc concepts that postulate effects that cancel each other so we cannot detect them (FitzGerald contraction). For Einstein, as for Kant, the idea of God expresses the idea of systematic unity and interconnectedness, as well as of simplicity, which Einstein often identifies with the unity and logical perfection of foundations. Of course, for both Kant and Einstein there is no sense in talking about a unilinear progress of theories that approximates correspondence with reality with ever-growing precision until the final and correct theory is reached. Still, the idea of "completed" science informs Einstein's philosophy and research—we proceed from a less "true" to a "truer" theory through further unification and systematization of a wider complex of phenomena.

As was mentioned earlier, Einstein held a "stratification" theory of the growth of knowledge (Einstein 1936, p. 294): Scientific knowledge advances from one of the levels, or "strata," to another by a process of the unification and simplification of foundations. In this process of systematization the domain of applicability of the theory becomes larger, its logical basis narrower, and the gap between the basic concepts of the theory and immediate experience wider. And even though, from the logical point of view, the previous, temporary stratum becomes superfluous after the next one is reached in this process, at any given time there is nevertheless only *one* adequate stratum that can serve as a springboard for the next one, until the final correct theory of the universe is reached.

The ideal of "total science" can explain what some scholars (Howard 1990) see as a peculiar blend of conventionalism and realism in Einstein's philosophy of science. Einstein believed that in principle there can be empirically equivalent alternative theories for any domain of phenomena.[8] Yet if for Duhem such an underdetermination of theory by facts means that there can be no ultimately correct theory, the Einsteinian approach in the Kantian spirit not only gives a meaning to the regulative ideal of a single, final, correct theory, but also illuminates Einstein's seemingly puzzling remarks that despite this underdetermination, at any given time there is only one correct theory: the theory with the greatest power of unification at that time. It is worthwhile to quote Einstein directly:

> In this state of methodological uncertainty one can think that arbitrarily many, in themselves equally justified, systems of theoretical physics were possible; and this opinion is, in principle, certainly correct. But the development of physics has shown that of all the conceivable theoretical constructions a single one has, at any given time, proved itself unconditionally superior to all others . . . furthermore this conceptual system which is unequivocally coordinated with the world of experience is reducible to a few basic laws from which the whole system can be developed. With every new important advance the researcher here sees his expectations surpassed, in that those basic laws are more and more simplified under the pressure of experience. With astonishment he sees apparent chaos resolved into a sublime order. (Einstein 1918, p. 226)

This ideal of total science—God's science—informs Einstein's belief that from studying the limited domain of theoretical physics one can reach the ultimate theory with "the proud name of the theory of the universe." Even though this ideal science is beyond the capacity of the human intellect, its unattainability is not "a matter of fundamental principle." This "total" science constitutes an epistemological ideal to which scientists must strive: "The supreme task of the physicist is to arrive at those universal elementary laws from which the cosmos can be built up by pure deduction" (Einstein 1918, p. 226).

The idea of the unity, or order, of Nature, as well as of God as a "heuristic fiction," is a very prominent theme in Stadler's lectures on Kant (Stadler 1912). Stadler discussed extensively the Kantian ideas of God, of causality, and of reality as absolute totality. The idea of God is theoretically empty—no empirical content can be associated with it. Because from the concept of an object we can never infer its existence, there can be no valid theological proof of the existence of God. Precisely for the same reason that we cannot infer God's existence, we cannot disprove it either. However, as Stadler emphasizes, the idea of a God with absolute wisdom and purposiveness can regulate reason in its progress, "as if" all the interconnections are the ordering of a supreme reason. Those researchers who confuse the regulative idea of God with an ontological idea make a very fruitful mistake: "Even an error can do no harm here. At most it can occasionally come to pass that, where we expect a teleological interconnection, there is merely a mechanical one" (Stadler 1912, p. 232). Stadler emphasizes this point forcefully: With the help of the belief in the unity and interconnectedness of Nature as springing from a first cause, a scientist can make "many useful scientific discoveries" ("*eine Menge nützliche Entdeckungen*").

The idea of meaningful interconnectedness underlies, in a quite direct way, not only Einstein's philosophy of science, but also the very process of his scientific work. More than anything else, it guides his actual scientific effort, from his early papers on quantum theory and relativity to the general theory of relativity and unified field theory. Where others saw lack of connection, or mere coincidence, Einstein presupposes a revelation of deep underlying lawfulness. The idea that there might be unnatural, complicated diversity and that God might have an "untidy" diffuse mind is "intolerable": "The idea that there exist two structures of space independent of each other, the metric gravitational and the electromagnetic, was intolerable to the theoretical spirit" (Einstein 1934a, p. 285).

While it is widely acknowledged that an urge for unification guided Einstein's scientific efforts from his work on general relativity onwards, the early Einstein is not usually presented in the literature as a unifier, but rather as a follower of the phenomenological and empiricist approach of Mach and Hume. Yet it is the search for systematic unity and interconnectedness that inspired in a central way Einstein's early scientific effort. While an extensive discussion of this theme is outside the scope of this paper, a few brief remarks are in place here.

The overarching search for unity in Einstein's early papers on the quantum theory is analyzed with rare intellectual empathy and insight in Martin Klein's pioneering work (Klein 1963, 1965, and especially Klein 1980). Einstein's attraction to the subject of the quantum theory was sparked by its unifying possibilities, by a connection that he investigated in his 1904 paper on the foundations of thermodynamics between the thermodynamic constant κ (essentially the Boltzmann constant, determined by the sizes of the elementary units of matter and electricity) and the order of magnitude of λ_m, the wavelengths at the energy maximum of radiation (Einstein 1904).

When Einstein returned to this subject in his 1905 paper on the quantum hypothesis, he began with the heart of what troubled him most—duality in the foundations of physics. This duality was felt most sharply in Lorentz's theory. The discrete mechanics of material bodies and the continuous field theory of electromagnetism employ very different sets of concepts and mathematical methods. Einstein found a striking similarity between the formula for the entropy of blackbody radiation in the Wien limit and the entropy of an ideal gas. It was this similarity that led to his hypothesis, or, as he very cautiously termed it at first, "heuristic point of view," that electromagnetic radiation in a certain (high frequency) region could also be viewed "as if" it consisted of a number of independent particles of energy. Einstein was cautious with the discovered analogy between matter and radiation. For he was not one to build hastily. And he immediately asked a deeper question: Is this analogy coincidental, meaningless, or does it point to something fundamental about the nature of radiation? (See Klein 1980.) Einstein analyzed the Boltzmann relation between entropy and probability, and came to the conclusion that it does not depend on any special assumption about the laws of motion of the molecules. That is why this expression is a general one and leads to the inevitable conclusion that, in the Wien limit, radiation behaves as if it had a corpuscular nature. The similarity between the fundamental concepts of matter and radiation is therefore not superficial; it points to some deep interconnectedness.

The same unifying spirit, which assumes that any meaningful scientific results must be subject to extension to other domains, characterizes Einstein's further work on quantum theory.[9] This led to Einstein's paper on the specific heats of solids (Einstein 1907), and to his laying the foundation for the quantum theory of matter.

Nor were Einstein's concerns in his quantum and relativity papers independent of each other. As Einstein himself pointed out in his Salzburg lecture in 1909, because the relativity paper abolished the ether, electromagnetic fields no longer have to be viewed as states of this hypothetical medium, and can be conceived as consisting of independent structures. Similarly, the equivalence of mass and energy, when light is exchanged between two bodies, means a transfer of mass between them, which is, again, akin to a corpuscular conception of light (Einstein 1909). I believe that these were not later retrospective realizations: Such interconnectedness guided, informed, and reinforced the very scientific efforts themselves.

If we turn to Einstein's 1905 relativity paper, it too begins with Einstein's pointing out the intolerable duality, or redundancy in the existing theoretical framework. This redundancy constitutes a striking contrast with the unity and simplicity of phenomena (Einstein 1905c).[10] The paper on relativity theory is in fact a paradigmatic example of a search for a unified system, with a minimum of independent basic assumptions and a maximum of consequences. The whole theory of special relativity is based entirely on two assertions, which Einstein elevated to the status of postulates:

1) The laws of physics take the same form in all inertial frames;
2) In a given inertial frame, the velocity of light is the same whether this light is emitted by a body at rest or in motion.

In order to achieve a reconciliation of these two seemingly contradictory postulates, Einstein introduced the epistemological analysis of the concepts of time and simultaneity with the help of considerations of measurement procedures and signal velocity. In positivistic circles, a lot of philosophical mileage was obtained later from this move by Einstein. Yet this move does not necessarily indicate a positivistic commitment on Einstein's part, for to ground the concept of time in its empirical verification and direct measurement procedure was a means, and not an end, for Einstein. The aim was to reconcile the two empirically supported propositions in order to secure a unified and simple foundation for physics. It is a singular heuristic move, not a demonstration of philosophical commitment.[11]

Einstein's demonstration of the reality of atoms (Einstein 1905b), his failure to mention the "crucial" Michelson-Morley experiments, and the strongly speculative components of his relativity paper (Einstein 1905c) all indicate that Einstein was hardly a committed Machian.[12] I would like to suggest that, despite the strongly phenomenalist and empiricist aspects of Einstein's early work, his overall scientific effort can hardly be labeled a positivistic one. In the case of the young Einstein, the seeds of Mach's influence fell on already existing rich Kantian soil.

4. Einstein's Realism and Determinism

I would like to conclude with a brief analysis of two central notions of Einstein's philosophical position from the Kantian perspective—his realism and determinism.

In his important book, Arthur Fine analyzed what he called Einstein's "unsystematic" remarks about realism and causality and came to the conclusion that Einstein's realism and causality are best understood through the notion of "entheorizing" (Fine 1984, p. 87). This term is used when the question whether such-and-so is the case is transformed into the question whether the theory in which such-and-so is the case is a viable theory. Thus, ontological questions about reality and causality are transformed into the question of the adequacy of scientific theories that are built on realistic or deterministic foundations.

Fine's characterization is a fitting one. As Einstein said about causality, "We speak of it when we have accepted a theory in which connections are represented as rational. . . . For us causal connections only exist as features of theoretical constructs" (Einstein to E. Zeisler, 10 December 1952, as quoted in Fine 1984, p. 87). Note the connection, even identity, between the concepts of "causality" and "rationality." Similarly, about reality: "It is basic for physics that one assumes a real world existing independently from any act of perception. But this we do not *know*. We take it only as a program of our scientific endeavors" (Einstein 1949, p. 674). What Fine calls "entheorizing" is very similar to Kant's central procedure of relocation of problems from an ontological to a methodological level. Yet there is also a difference: The notion of "entheorizing" is stripped of any metaphysical underpinnings. However, for both Kant and Einstein the metaphysical function of realism and causality was essential.[13] For Einstein, both realism and determinism, which he identified with rationality, were the very presuppositions of scientific research. Without those notions the very concept of science and scientific research becomes meaningless. As for Kant, so for Einstein, realism and causality were the transcendental principles of science.

In complete agreement with the Kantian approach, which aims to insulate the basic notions of philosophy from their ontological underpinnings, Einstein often states that the question of scientific realism is meaningless outside the domain of a unified conceptual model. In his "Reply to Criticisms" there are explicitly Kantian remarks: "The real is not given to us, but put to us (by way of a riddle)." Einstein explained:

> There is a conceptual model . . . whose authority lies solely in its verification. This conceptual model refers precisely to the "real" (by definition), and every further question concerning 'the nature of the real' appears empty. (Einstein 1949, p. 680)

It is clear that Einstein's realism is very different from that of the "correspondence theory of truth." Einstein rejects the Cartesian image of theoretical construction as guessing the workings of a closed watch: The scientist "*will never be able to compare his picture with the real mechanism, and he cannot even imagine the possibility or the meaning of such a comparison*" (as quoted in Fine 1984, p. 93). There is a wonderful statement in a letter of Einstein to the Bonn mathematician, Eduard Study, who defended a version of scientific realism as correspondence with reality: "'The world of bodies is real'. . . . The above statement appears to me to be, in itself, meaningless, as if one said: 'the world of bodies is cock-a-doodle-doo.' It seems to me that 'real' is in *itself* an empty, meaningless category" (quoted in Howard 1990, p. 368).

It is remarks of this kind that have led Einstein scholars to question whether Einstein was, after all, a realist. In Arthur Fine's paper on "Einstein's Realism," there is a section entitled: "Was Einstein a Realist?"

(Fine 1984, pp. 105–109), and Don Howard called his recent paper, "Was Einstein Really a Realist?" (Howard 1993).

Howard identifies Einstein as a conventionalist, while Fine characterizes Einstein's position as being close to van Fraassen's "constructive empiricism." Both of these characterizations are only partially true. Einstein's criteria for truth, objectivity, and reality are holistic in the Kantian sense: Empirical adequacy is merely a necessary, but not a sufficient condition for scientific truth. I cannot agree therefore with Arthur Fine's conclusion that, for Einstein, realism is merely a motivation, an emotional drive, and not an epistemological stance (Fine 1984, pp. 109–111).

It is true that many of Einstein's remarks about his search for truth are very emotional. Kant, himself, in his *Critique of Judgement*, wrote about the great emotional satisfaction of scientific research: "The discovery that two or more empirical heterogeneous laws of nature may be combined under one principle comprehending them both, is the ground of very marked *pleasure*, often even of admiration" (quoted in Buchdahl 1969, p. 498).

Einstein's language is much more passionate: He compares the drive for scientific research to the state of mind of the religious worshiper, or a lover's infatuation (Einstein 1918, p. 227). There is no question that science provided for Einstein an escape from the uncontrollability of everyday life and that comprehending the deepest secrets of creation must have served not only as the source of positive self-identity, but of a feeling of enormous power and of proximity to God. In 1929, in his essay "Über den gegenwärtigen Stand der Feldtheorie," Einstein wrote about his urge to understand not only the "how," but the "why," of Nature's ways, comprehending the essence of the necessity of those connections that, in Einstein's words, God himself could not have arranged otherwise: "This is the *Promethean* element of scientific experience . . . the particular magic of scientific considerations" (as quoted in Holton 1968, p. 260; emphasis added).

Yet such expressions do not indicate that Einstein's realism is merely motivational. The uniqueness of Einstein's epistemological position in a Kantian spirit lies precisely in its avoidance of the pitfalls of both extreme empiricism and naive realism.[14]

We can now perhaps better understand Einstein's position on causality. As Arthur Fine pointed out, causality is central to Einstein's realism (Fine 1984, p. 97). In many of Einstein's remarks on the subject, reality and causality, or strict lawfulness, are organically linked: "Physics is an attempt at the conceptual construction of a model of a real world, as well as its lawful structure" (Einstein to Schlick, 28 November 1930, as quoted in Fine 1984, p. 97). Again, Einstein's adherence to causality does not represent merely an emotional attachment, or an inexplicable metaphysical belief. From a Kantian perspective, based on the conception of "systematic unity," reality and causality, objectivity and strict lawfulness are equivalent notions. In a Kantian framework, a thing's existence is inferred from the fact that it has a place in the connected whole of our experience—that between it and other things a connection according to a law exists. This aspect was emphasized in Stadler's lectures that Einstein attended as a student. Of course, there is no logical connection in general between realism and determinism. We can, for example, have realistic acausal theories (kinetic theory of gases). Yet in the Kantian system, unity, simplicity, interconnectedness, comprehensibility, rationality, reality, and causality are all almost synonymous and reach back to the "as if" conception of God. Einstein's sharing of Kant's basic presuppositions explains his epistemic (not merely emotional) commitment to causality, as well as the fact that Einstein's concept of causality was an indispensable aspect of his realism. This position makes sense only in the Kantian framework. In alternative frameworks, such as positivism or naive realism, the necessary connection between realism and causality does not make sense.

5. Conclusion

I have argued in this paper that, from a Kantian perspective, we can gain a more coherent picture of Einstein's philosophy of science and obtain an epistemological legitimization of Einstein's position that otherwise looks quite arbitrary. The Kantian framework allows us to find a meaningful interdependence of different aspects of Einstein's approach that otherwise seem disconnected.

The realization that Einstein's philosophy of science has prominent Kantian components perhaps should not come as a surprise. Our understanding of Einstein's philosophical legacy relied in the past strongly —perhaps too strongly—on the writings of logical positivists who dissociated themselves from Kant and appropriated Einstein's achievements for their own aims. Yet against the background of recent scholarship that has uncovered significant Kantian and neo-Kantian components in the writings of logical positivists (Friedman 1983, 1987; Ryckman 1991; Coffa 1991), Kantian underpinnings of Einstein's thought can hardly be seen as an aberration.

Einstein's relation to Kant did not remain unchanged during his life. In the 1920s, when Einstein waged his battle with neo-Kantians over the fundamental status of the Euclidean-Kantian framework, his remarks about Kant are often lukewarm, while those about neo-Kantians themselves are just blunt. In later years, Einstein's tone changes to one of deep appreciation of the Kantian legacy. Clearly, a future study of Kant's impact on Einstein will have to deal with the broader, changing, socio-cultural context, as well as with the detailed analysis of Einstein's concrete scientific effort.

Acknowledgements. I am grateful to John Stachel and Don Howard for helpful discussions and their encouragement to pursue this study. I would also like to express my gratitude to my friends Annette and Nick Gmür for their help with Stadler's Nachlass.

This paper was written before Michael Friedman's important book *Kant and the Exact Sciences* (Cambridge, Massachusetts: Harvard University Press, 1992) became available to me.

The research for this paper was supported in part by National Science Foundation Grant DIR-9011053.

NOTES

[1] The pioneering studies of Einstein's philosophical make-up are those of Gerald Holton, collected in Holton 1988. Philosophical influences on Einstein are extensively explored also in Miller 1981, in Goldberg 1967, 1970, and in a series of recent papers by Don Howard (1994, 1990, 1984).

[2] On the wide dissemination and impact of Kant at the turn of the century, see Willey 1978, and on teaching and research in Kantian philosophy in German universities, see Köhnke 1991.

[3] Thus, Einstein wrote to Freud around 1931: "This great aim [of liberation of mankind from war—M.B.] has been professed by all those who have been venerated as moral and spiritual leaders beyond the limits of their own time and country without exception, from Jesus Christ to Goethe and Kant" (Einstein 1974, p. 104; the letter was originally published in Einstein 1934b).

[4] Platter acknowledges in his foreword to Stadler 1912 that the book's contents are taken from Stadler's *Nachlass*.

[5] While Einstein's references to Hume's impact are often treated as a token of Einstein's empiricism, the discussion in this paper leads to a different appreciation of Hume's role in Einstein's critical thinking: "Man has an intense desire for assured knowledge. That is why Hume's clear message seemed crushing: the sensory raw material, the only source of our knowledge, through habit may lead us to belief and expectation but not to the knowledge and still less to the understanding of lawful relations" (Einstein, 1944, p. 22). Hume's impact on Einstein is then not

necessarily an influence in the direction of empiricism, as usually assumed.

[6] Arthur Fine, impressed by Einstein's unfailing emphasis on the empirical component of our knowledge, reached the conclusion that Einstein's philosophical position is close to that of the "constructive empiricism" of van Fraassen (Fine 1984, p. 108).

[7] For the realization of the importance of this idea in Kant's philosophy of science I am indebted to Gerd Buchdahl (1969), whose interpretation I follow closely.

[8] This view is expressed, for example, in the paper Einstein published in the *Berliner Tageblatt* in 1919: "If two theories are available, both of which are compatible with the given factual material, then there is no other criterion for preferring the one or the other than the intuitive view of the researcher. Thus we may understand how sharp-witted researchers, who have command of theories and facts, can still be passionate supporters of contradictory theories" (quoted in Howard 1990).

[9] If Planck's hypothesis "strikes to the heart of the matter," then the limitation of the energy of oscillators to discrete energy values should be evident not only in the interaction of matter and radiation, but also in other areas, and one must find that a discrepancy should arise between experiment and the classical kinetic-molecular theory in other areas of the theory of heat.

[10] In classical electrodynamics there exist two independent explanations for the phenomena where magnet and conductor are in relative motion. If the magnet is in motion, there arises an electric field in the vicinity of the conductor, producing current. When the conductor is in motion and the magnet at rest, there is a totally different explanation, based on the concept of electromotive force (Holton 1988; Miller 1981). Examples of this sort, Einstein explains, were the motive for his search for a more unified and coherent foundation for physics.

[11] Poincaré himself, who hardly can be accused of positivism and empiricism, questioned the notion of absolute simultaneity a few years earlier, and recommended an approach to this problem by direct measurement procedures, based on the transmission of light, this in an essay that Einstein had most likely read. Poincaré's viewpoint, as compared with Einstein's, is analyzed in a most penetrating way in Pais 1982.

[12] Einstein's belief in the reality of atoms, as opposed to light quanta, which constituted initially only a "heuristic point of view," is consistent with the principle of the "systematic unity of nature" as a criterion for theoretical truth. The atomic conception of matter was embedded in a wide theoretical framework of physics and chemistry, while the atomic conception of light had yet to prove its mettle. About Einstein's relation to the Michelson-Morley experiments, see Holton 1988, ch. 8.

The antipositivistic components of Einstein's relativity paper were perceptively analyzed in Holton 1968. Moreover, at least one component that Holton considered as a Machian one (Einstein's "insistence that the fundamental problems of physics cannot be understood until an epistemological analysis is carried out") is at least as much Kantian, as it is Machian.

[13] It is questionable that Einstein was "really" ready to subject the questions of "realism" and "causality" to a test of the success of competing research programs: those programs that comprise the realistic-deterministic theories vs. programs built on realistic and acausal foundations. Einstein's relation to quantum theory suggests that his view that one is not "forced to cling a priori" to the program of realism was only a gesture of openness on this issue in principle, not in practice.

[14] This advantage of Kant's approach is argued forcefully in Kitcher 1986.

References

Born, Max (1971). *The Born-Einstein Letters: Correspondence between Albert Einstein and Max and Hedwig Born from 1916 to 1955*. I. Born, trans. New York: Walker & Company.

Buchdahl, Gerd (1969). *Metaphysics and the Philosophy of Science. The Classical Origins: Descartes to Kant*. Cambridge, Massachusetts: MIT Press.

Butts, Robert E., ed. (1986). *Kant's Philosophy of Physical Science*. Dordrecht: D. Reidel.

Coffa, J. A. (1991). *The Semantic Tradition from Kant to Carnap: To the Vienna Station*. Linda Wessels, ed. Cambridge, UK: Cambridge University Press.

Duhem, Pierre (1906). *La Théorie physique: son objet et sa structure*. Paris: Chevalier & Rivière.

Einstein, Albert (1904). "Zur allgemeinen molekularen Theorie der Wärme." *Annalen der Physik* 14: 354–362.

— (1905a). "Über einen die Erzeugung und Verwandlung des Lichtes betreffenden heuristischen Gesichtspunkt." *Annalen der Physik* 17: 132–148.

— (1905b). "Über die von der molekularkinetischen Theorie der Wärme geforderte Bewegung von in ruhenden Flüssigkeiten suspendierten Teilchen." *Annalen der Physik* 17: 549–560.

— (1905c). "Zur Elektrodynamik bewegter Körper." *Annalen der Physik* 17: 891–921.

— (1907). "Die Plancksche Theorie der Strahlung und die Theorie der spezifischen Wärme." *Annalen der Physik* 22: 189–90, 800.

— (1909). "Über die Entwicklung unserer Anschauungen über das Wesen und die Konstitution der Strahlung." *Physikalische Zeitschrift* 10: 817–825.

— (1914). "Antrittsrede." *Königlich Preussische Akademie der Wissenschaften* (Berlin). *Sitzungsberichte*: 739–742. Page numbers and quotations taken from the translation as "Principles of Theoretical Physics." In Einstein 1974, pp. 220–223.

— (1918). "Motive des Forschens." In *Zu Max Plancks sechszigstem Geburtstag. Ansprachen, gehalten am 26. April 1918 in der Deutschen Physikalischen Gesellschaft*. Karlsruhe: Müller, pp. 29–32. Page numbers and quotations taken from the translation as "Principles of Research." In Einstein 1974, pp. 224–227.

— (1933). *On the Method of Theoretical Physics.* The Herbert Spencer Lecture, delivered at Oxford, June 10, 1933. Oxford: Clarendon. Page numbers taken from the reprinting in Einstein 1974, pp. 270–276.

— (1934a). "Das Raum-, Äther- und Feld-Problem in der Physik." In Einstein 1934b, pp. 229–248. Page numbers and quotations taken from the translation as "The Problem of Space, Ether, and the Field in Physics." In Einstein 1974, pp. 276–285.

— (1934b). *Mein Weltbild.* Amsterdam: Querido Verlag.

— (1936). "Physics and Reality." *Journal of the Franklin Institute* 221: 349–382. Page numbers taken from the reprinting in Einstein 1974, pp. 290–323.

— (1944). "Remarks on Bertrand Russell's Theory of Knowledge." In *The Philosophy of Bertrand Russell.* Paul Arthur Schilpp, ed. Evanston, Illinois: The Library of Living Philosophers, pp. 278–292. Page numbers taken from the reprinting in Einstein 1974, pp. 18–25.

— (1946). "Autobiographical Notes." In Schilpp 1949, pp. 1–94.

— (1949). "Remarks Concerning the Essays Brought Together in This Cooperative Volume." In Schilpp 1949, pp. 665–688.

— (1974). *Ideas and Opinions.* New York: Crown.

— (1987). *The Collected Papers of Albert Einstein.* Vol. 1, *The Early Years, 1879–1902.* John Stachel et al., eds. Princeton: Princeton University Press.

Fine, Arthur (1984). "Einstein's Realism." In *Science and Reality: Recent Work in the Philosophy of Science, Essays in Honor of Ernan McMullin.* James T. Cushing , Cornelius F. Delaney, and Gary Gutting, eds. Notre Dame, Indiana: University of Notre Dame Press, pp. 106–133. Page numbers taken from the reprinting in *The Shaky Game: Einstein, Realism and the Quantum Theory.* Chicago: University of Chicago Press, 1986, pp. 86–111.

Friedman, Michael (1983). "Critical Notice: Moritz Schlick, Philosophical Papers." *Philosophy of Science* 50: 498–514.

— (1987). "Carnap's Aufbau Reconsidered." *Nous* 21: 521–545.

Goldberg, Stanley (1970). "Poincaré's Silence and Einstein's Relativity: The Role of Theory and Experiment in Poincaré's Physics." *British Journal for the History of Science* 5: 73–84.

Holton, Gerald (1968). "Mach, Einstein, and the Search for Reality." *Daedalus* 97: 636–673. Page numbers taken from the reprinting in Holton, 1988, pp. 237–277.

— (1988). *Thematic Origins of Scientific Thought: Kepler to Einstein,* rev. ed. Cambridge, Massachusetts: Harvard University Press.

Howard, Don A. (1984). "Realism and Conventionalism in Einstein's Philosophy of Science: The Einstein-Schlick Correspondence." *Philosophia Naturalis* 21: 618–29.

— (1990). "Einstein and Duhem." In *Pierre Duhem: Historian and Philosopher of Science.* Roger Ariev and Peter Barker, eds. *Synthese* 83: 363–384.

— (1993). "Was Einstein Really a Realist?" *Perspectives on Science* 1: 204–251

— (1994). "Einstein, Kant, and the Origins of Logical Positivism." In *Language, Logic and the Structure of Scientific Theories: The Carnap-Reichenbach*

Centennial. Wesley Salmon and Gereon Wolters, eds. Pittsburgh: University of Pittsburgh Press, pp. 45–105.

Kant, Immanuel (1787). *Critique of Pure Reason*, 2nd ed. Norman Kemp Smith, trans. London: Macmillan & Co., 1929. Rpt. New York: St. Martin's Press, 1965.

Kitcher, Philip (1986). "Projecting the Order of Nature." In Butts 1986, pp. 201–235.

Klein, Martin J. (1963). "Einstein's First Paper on Quanta." *Natural Philosopher* 2: 57–86.

— (1965). "Einstein, Specific Heats, and the Early Quantum Theory." *Science* 148: 173–180.

— (1980). "No Firm Foundation: Einstein and the Early Quantum Theory." In *Some Strangeness in the Proportion: A Centennial Symposium to Celebrate the Achievements of Albert Einstein.* Harry Woolf, ed. Reading, Massachusetts: Addison-Wesley, pp. 161–186.

Köhnke, Klaus (1991). *The Rise of Neo-Kantianism: German Academic Philosophy between Idealism and Positivism.* R.J. Hollingdale, trans. Cambridge, UK: Cambridge University Press, 1991.

Miller, Arthur I. (1981). *Albert Einstein's Special Theory of Relativity.* London: Addison-Wesley.

Pais, Abraham (1982). *"Subtle is the Lord . . ." The Science and the Life of Albert Einstein.* Oxford: Oxford University Press.

Ryckman, T. A. (1991). "Conditio sine qua non? Zuordnung in the early epistemologies of Cassirer and Schlick." *Synthese* 88: 57–95.

Schilpp, Paul Arthuer, ed. (1949). *Albert Einstein: Philosopher-Scientist.* Evanston, Illinois: The Library of Living Philosophers.

Stadler, August (1912). *Kant. Akademische Vorlesungen.* Leipzig: Johann Ambrosius Barth.

Willey, Thomas E. (1978). *Back to Kant: The Revival of Kantianism in German Social and Historical Thought, 1860–1914.* Detroit: Wayne State University Press.

Einstein's Controversy with Drude and the Origin of Statistical Mechanics: A New Glimpse from the "Love Letters"

Jürgen Renn

1. Introduction: Statistical Mechanics as a Conduit from Classical to Modern Physics

In this work I argue that statistical mechanics, at least in the version published by Einstein in 1902 (Einstein 1902b), was the result of a reinterpretation of already existing results by Boltzmann. I will show that, for this reinterpretation, a certain perspective on these results was decisive, a perspective shaped by Einstein's concern with specific problems of the constitution of matter and radiation, as well as with atomism as a general foundation for physics. Using newly-available evidence, I will identify the electron theory of metals as the key problem triggering the elaboration of statistical mechanics. In this way, a conjecture by the Ehrenfests on the role of electron theory for a renewal of statistical physics, as well as a hypothesis by the editors of Volume 2 of Einstein's *Collected Papers* concerning its role for Einstein's work, receive an unexpected confirmation. In addition, I will argue that a controversy between Einstein and Paul Drude in 1901 was, in effect, not so much a dispute about the latter's electron theory of metals, as has been assumed so far, but a controversy in which Einstein's true opponent was, at least in part, Boltzmann and whose issue was the foundation of an atomistic theory of matter. This controversy became the starting point for Einstein's elaboration of his own approach to statistical mechanics.

The development of statistical physics in the 19th century was closely associated with the hope to extend the principles of mechanics to the range of thermal phenomena and to establish atomism as a conceptual foundation of physics. The work of Maxwell and Boltzmann on the kinetic theory of gases provided a solid foundation for this hope. It was based on an idealized but nevertheless rather concrete model of a gas as a collection of

Einstein Studies, vol. 8: Einstein: The Formative Years, pp. 107–157.

particles moving freely through an essentially empty space and occasionally interacting with each other. By means of these interactions, it could be shown, kinetic energy would become equally distributed through the physical system, resulting in a statistical "equipartition of energy," which characterizes the state of thermal equilibrium. Many features of macroscopic systems described by phenomenological thermodynamics could be reconstructed within the kinetic theory of gases. In addition, the kinetic theory allows for a number of rather surprising statistical assertions concerning the atomistic constituents of a gas, such as the distribution of the energies of the single particles known as the "Maxwell-Boltzmann distribution."[1]

Although 19th century kinetic theory was a means to extend the range of the mechanical foundation of physics, 20th century statistical mechanics became a means to question and eventually to overcome just these foundations. Arguments based on statistical mechanics (concerning, in particular, the equipartition of energy) made it possible to show that classical electrodynamics was unable to cope with the thermal equilibrium of heat radiation. Among the first to attain this insight into the limits of classical radiation theory was Albert Einstein in 1905 (Einstein 1905a). Two years later he applied methods of statistical mechanics in a first attempt at a quantum theory of solid bodies (Einstein 1907).

Since statistical mechanics does not require us to follow the time development of the interaction between individual atomistic constituents of a macroscopic system, it is much more generally applicable than the kinetic theory of gases. Statistical mechanics, rather, considers the statistical properties of a "virtual" ensemble of macroscopic systems, all of which are characterized by the same dynamics but which vary in the initial values of their atomistic constituents. Different kinds of thermodynamical systems in equilibrium are represented by different such statistical ensembles—an isolated thermodynamic system by a "microcanonical ensemble," in which all members have the same (or approximately the same) energy: a system that is not isolated but is in contact with a heat reservoir held at a fixed temperature by a "canonical ensemble," in which the energies of the members obey a certain statistical distribution (characterized by an exponential function) allowing them to take all possible values. Due to its generality, statistical mechanics can be employed in classical and, with appropriate modifications, also in quantum physics. It is for this reason that it could play a key role in the transition from classical to modern quantum physics.

Building blocks of statistical mechanics can be found already in the numerous publications of the eminent protagonists of the 19th century kinetic theory of heat, James Clerk Maxwell and Ludwig Boltzmann. It was, however, only in the book that Josiah Williard Gibbs published in 1902, entitled *Elementary Principles in Statistical Mechanics* (Gibbs 1902), that statistical mechanics was given, for the first time, a coherent and autonomous formulation. The now standard terminology "microcanonical ensemble" and "canonical ensemble" for two essential concepts of statistical mechanics is due to him. In the same year, 1902, Einstein published the first of a series of three papers on statistical physics, entitled *Kinetic Theory of Thermal Equilibrium and of the Second Law of Thermodynamics* (1902), *A Theory of the Foundations of Thermodynamics* (1903), and *On the General Molecular Theory of Heat* (1904) (Einstein 1902b, 1903, 1904). These three papers are based on Boltzmann's major book on the kinetic theory of heat, entitled *Vorlesungen über Gastheorie* (Boltzmann 1896, 1898). But they established, independently of Gibbs, statistical mechanics and provided the basis for Einstein's exploration of the consequences of the quantum hypothesis for a revision of the foundations of classical physics, and also for his analysis of Brownian motion and other fluctuation phenomena as evidence for the existence of atoms (Einstein 1905c; for an historical discussion, see Einstein 1989, the editorial note "Einstein on Brownian Motion," pp. 206–222).

There can be little doubt that statistical mechanics, with its important impact on the further development of 20th century physics, constitutes a conceptual innovation in the history of science. The fact that so many of its building blocks are found in the work of Maxwell and Boltzmann, and hence actually predate its creation, suggests that this innovation was largely due to a change of perspective, to a reinterpretation of preexisting results in a new light. In Einstein's case, this new light was, as I will show in detail, provided by a new context of application of statistical physics.

That a certain revitalization of statistical physics at the beginning of the new century was caused by a new context of application is a conjecture that was first expressed by Paul and Tatiana Ehrenfest in 1911. They also point, in particular, to the electron theory of metals as one such new context:

> In particular, the last few years have seen a sudden and wide dissemination of Boltzmann's ideas (the H-theorem, the Maxwell-Boltzmann distribution, the equipartition of energy, the relationship between entropy and probability, etc.). However, one cannot point at a corresponding progress in the conceptual clarification of Boltzmann's system to which one can ascribe this turn of affairs.
>
> It is much more likely that the study of electrons and the investigation of colloidal solutions with the ultramicroscope have been responsible. In general,

> both of these have had the effect of reviving and deepening the concept that all bodies can be pictured as aggregates of a finite number of very small and identical elementary components, and that correspondingly every process in a physical or chemical problem which can be observed by normal methods is a complex of an enormously large number of individual processes. The opportunity arose to apply the methods of the kinetic theory of gases to completely different branches of physics. Above all, the theory was applied to the motion of electrons in metals (V 14, Section 40, by H.A. Lorentz), to the Brownian motion of microscopically small particles in suspensions (Section 25), and to the theory of blackbody radiation (V 23, by W. Wien). (Ehrenfest and Ehrenfest 1959/1990, p. 68; the parenthetical references are to other sections of the *Encyclopädie*)

The impact of novel applications on a conceptual system is, however, not necessarily limited to such a revitalization. New circumstances of application can change the meaning of concepts, and attempts to solve new problems by traditional conceptual means may lead to shifts of emphasis within the conceptual system (see the introduction to Damerow et al. 1992). What was a marginal and problematic result from one point of view, that is, as interpreted within one conceptual system, may come to constitute the core of another, new conceptual system. For example, the result, established by Michelson, Morley and others towards the end of the 19th century, that no effect of the earth's motion with respect to the hypothetical ether can be found was, in this way, transformed from a stumbling block of Lorentz's ether-based electrodynamics into the cornerstone of Einstein's special theory of relativity, in the form of the principle of relativity.[2]

At this point, a more general theoretical observation may be in place: Such a continuous transition from one conceptual system to another, different one, is possible in a science like physics because its conceptual systems are formulated in a controlled technical language which displays characteristics of both a natural language and a formal system. While such a conceptual system is flexible enough to cope with a wide range of experience, it is also rigid enough occasionally to display inconsistencies when separate legitimate applications of a concept lead to extensions of meaning that turn out to be incompatible. The representation of physical results in terms of language and mathematical formalism provides an important instance of mediation between the old and the new system. A new conceptual system can emerge from reinterpreting a mathematical representation of physical concepts and their relation, the Lorentz transformations, for example, no longer seen as depending (albeit in a problematic way) on the old conceptual system, but rather as defining the relation between fundamental concepts of the new system, here the relativistic concepts of space and time.

In this paper, I will argue that a similar process of reinterpretation explains the emergence of statistical mechanics at the hands of Einstein. In other words, Boltzmann's work on the kinetic theory of heat played an analogous role in Einstein's creation of statistical mechanics to that of Lorentz's work on the electrodynamics of moving bodies in his creation of special relativity. To take one example that I will consider in some detail below, the equipartition of energy for arbitrary material bodies in thermal equilibrium was only a marginal topic in Boltzmann's work. He did assemble the mathematical tools necessary for its derivation in the sense of the later statistical mechanics, but actually considered such a demonstration to be problematic because of its dependence on a far-going hypothesis about the dynamical properties of arbitrary bodies.[3] To Boltzmann such a derivation had to be strictly grounded in the principles of mechanics, with the consequence that he placed at the center of his research those physical systems that he considered tractable under this condition, in particular gases. He also introduced the virtual statistical ensembles that later became core concepts of statistical mechanics, but he never systematically analyzed their relation to each other, although he had developed the means for such an analysis as well.

New contexts of application, in particular the electron theory of metals and the theory of heat radiation, shifted the emphasis within the kinetic theory of heat, as it was perceived by Einstein, from Boltzmann's questions concerning the mechanical foundations of the theory of heat to the problem of the derivation of the equipartition law for general physical systems in thermal equilibrium, as well as to related problems.[4] This shift of emphasis brought him to identify, as I will show in detail below, a "gap" in Boltzmann's work. Instead of solving the questions left by Boltzmann, e.g., concerning the dynamical properties of general physical systems, Einstein introduced new ones and reassembled Boltzmann's results in order to fill the "gap" he perceived.

This change of perspective had the consequence that these results now assumed a new meaning as cornerstones of a new approach: statistical mechanics. The relation between canonical and microcanonical ensembles, for instance, which Boltzmann had touched upon only in passing, became a key representation of the relation between a physical system at a fixed temperature and a heat reservoir. On the basis of this relation, Einstein was able to derive not only the equipartition theorem for more general physical systems at a given temperature but also genuinely novel results, as for instance a formula for energy fluctuations of such systems. One can indeed characterize Einstein's turn of perspective with similar words to those used by Martin Klein in order to explain Gibbs' indifference with regard to the discussion about irreversibility which had been so important to Boltzmann's work; commenting on a pertinent observation by Ehrenfest, Klein remarked:

> He was quite correct in saying that Gibbs had largely ignored almost all issues over which this battle had raged. But Gibbs took a very different view of the structure of the subject from Ehrenfest, or Boltzmann, and he was not trying to solve the same problem. (Klein 1970, p. 129)

In Einstein's case we are in a position to analyze in detail how such a very different view emerged out of his concerns with specific problems related to an atomistic theory of matter and to the theory of radiation. In order to reconstruct the emergence of his perspective, I will first discuss the evidence that has until now been available on his early scientific interests. This evidence suggests that Einstein's interest in the electron theory of metals may have been one of several research topics motivating his search for a generalization of the kinetic theory of gases (section 2). A reconstruction of a controversy he had with Drude concerning the electron theory of metals will not only confirm this influence but also show that this controversy directly triggered the elaboration of statistical mechanics.

In the light of newly available contemporary evidence, I will show in particular that this controversy, whose content was until now unknown, concerned, at least in part, Boltzmann's statistical physics and also affected an early attempt by Einstein to obtain a doctorate (section 3). The conceptual bond connecting Einstein's various early scientific interests, such as that in the electron theory of metals, is then characterized as a kind of "interdisciplinary atomism" that was implicit in turn-of-the century-physics but that was not generally pursued as a systematic research program.

As an illustration for the impact on Einstein's thinking of the links between his different research interests, I will offer a tentative reconstruction of his decision to reject his own approach to an electron theory of metals (section 4). Einstein's quest for an interdisciplinary atomism is shown to be the presupposition also of his identification of the "gap" in Boltzmann's work on the kinetic theory of heat, a gap whose very existence was denied by both Boltzmann and Drude. Einstein's atomism is then shown to constitute the perspective from which he could reinterpret elements of Boltzmann's work as building blocks of statistical mechanics, the new approach capable of filling this gap (section 5). Finally, in an epilogue, the contrast between the original intentions behind Einstein's work on statistical mechanics and its eventual consequences is discussed with the aim of removing the teleological aura surrounding this work, which is evoked by its later, revolutionary consequences for 20th century physics (section 6).

2. Einstein's Early Scientific Activities according to the "Love Letters"

Until a few years ago very little was known about Einstein's scientific interests and activities prior to his path-breaking papers of 1905 on the light quantum hypothesis, Brownian motion, and the electrodynamics of moving bodies. Apart from a handful of contemporary letters, the prehistory of these papers could only be reconstructed on the basis of Einstein's few earlier publications and of his later recollections,. The situation changed considerably when his correspondence with Mileva Marić, beginning in 1897, (the so-called "love letters") was published in the first volume of the *Collected Papers of Albert Einstein* (Einstein 1987; for an English translation of the "love letters," see Renn and Schulmann 1992). In addition to insights into the dramatic relation between Mileva Marić and Albert Einstein, these letters revealed several hitherto unknown and rather diverse scientific interests of the young Einstein, e.g., in radiation experiments and in the electron theory of metals; they show him to be an avid reader of textbooks and of contemporary physics journals; they bear witness to his early contacts with leading physicists such as Boltzmann, Drude, and Wien; and last but not least, they contain important hints concerning the prehistory of his 1905 papers.

Elsewhere I have claimed, on the basis of an analysis of Einstein's letters, that one can recognize a close relation between these hints and Einstein's other, seemingly unrelated scientific interests in this period, such as the electron theory of metals (Renn 1993; see also the introduction to Renn and Schulmann 1992, and in Einstein 1989, the editorial notes "Einstein's Dissertation on the Determination of Molecular Dimensions," pp. 170–182, and "Einstein on Brownian Motion," pp. 206–222). According to this reconstruction, key ideas of the papers of his *annus mirabilis* 1905 had first been developed under the auspices of these seemingly unrelated physical interests, which were above all characterized, according to this interpretation, by Einstein's search for a conceptual unity of physical phenomena on an atomistic foundation. It was this search that gave intellectual coherence to his early scientific endeavors, ranging from thermoelectricity, via radiation theory, to the electrodynamics of moving bodies, long before it turned out that the outcome of the search would shake the conceptual foundation of classical physics.

But in spite of the availability of Einstein's "love letters" as a new source for reconstructing his early intellectual development, the evidence remains slim. Many of the letters merely contain allusions to conversations with Marić about scientific subjects or brief sketches of Einstein's ideas, from which their content can hardly be reconstructed; other letters only mention an argument, without giving it. The years 1901 and 1902, for instance, are characterized by controversies, but each controversy represents a riddle. During this time, Einstein unsuccessfully attempted to obtain a doctorate, first under one supervisor, with whom he had a falling-out for unknown reasons, then with another one who rejected, for unknown reasons, a dissertation by Einstein whose precise content has also remained obscure.[5]

In between these two controversies he had, in mid-1901, a dispute with Paul Drude concerning the electron theory of metals, one of the previously unknown research interests revealed by the "love letters." In analogy to the freely moving molecules of a gas, Drude's theory assumes freely moving charge carriers inside a metal, accounting both for its electric and its thermal conductivity, as well as for the connection between these conductivities described by the Wiedemann-Franz law. The nature of Einstein's objections to Drude's theory, which was published in 1900 (Drude 1900a and b; see also Drude 1902), has remained just as unknown to us as the character of Drude's response to a letter we know Einstein had written to him in June 1901 (see Einstein to Mileva Marić, 4 June 1901, Einstein 1987, Doc. 112, p. 306). There is a puzzling hint at a possible connection between Einstein's dissertation and his controversy with Drude, but it has remained unclear whether this connection was one of content or was just due to Einstein's spirit of rebellion which pervaded all of these controversies (Einstein to Mileva Marić, 17 December 1901, Einstein 1987, Doc. 128, p. 326). I will return below to this riddle.

The obscurity surrounding Einstein's controversy with Drude is particularly unfortunate given its potential significance for understanding Einstein's development, both intellectual and personal. What is known about the controversy provides some insights into the way in which he took on the role of a rebel against scientific authorities, as well as into what appears to have been, at the time, a central focus of his scientific interests. The correspondence with Marić suggests, in fact, that Einstein had developed, independently of Drude, his own approach to an electron theory of metals (Einstein to Mileva Marić, 4 April 1901, Einstein 1987, Doc. 96, pp. 284–285). Hints at a relation between Einstein's interests in electron theory and in statistical physics have given rise to an interesting suggestion concerning the role of the electron theory of metals for the emergence of his statistical mechanics, as one among several of his contemporary research interests:

> Aside from his reading of material directly concerned with kinetic theory, Einstein was studying at least three other topics in 1901 and 1902 that may have suggested the need to extend the foundations of thermodynamics and kinetic theory. First, since at least the spring of 1901, Einstein had been reading Planck's papers on irreversible radiation processes, in which Planck sought to extend the concept of entropy to radiation. Second, Einstein was studying the work of Drude and others on the electron theory of metals, in which the apparatus of kinetic theory is employed to explain such phenomena as electric and thermal conductivity. . . . Third and most important, Einstein was working on a theory of molecular forces. (The editorial note in Einstein 1989, "Einstein on the Foundations of Statistical Physics," pp. 45–46; see also the discussion in Renn 1993, pp. 332–334)

The surmised link between Einstein's interest in the electron theory of metals and his elaboration of statistical mechanics becomes particularly plausible in view of evidence for his recognition of the central role of the equipartition of energy in such a theory, as we will discuss in detail below. But on the basis of the historical sources so far available, there was nevertheless little possibility of coming to definite conclusions about the complex relations between Einstein's interest in electron theory, his controversy with Drude, his failed dissertation, and his creation of statistical mechanics.

Under these circumstances it deserves attention that a previously unknown passage from a letter by Mileva Marić to Albert Einstein, concerning Drude's response to Einstein's objections, has now come to light. The passage is contained in a letter of which only a partial copy was available to the editors of Vol. 1 of the *Collected Papers of Albert Einstein*. The complete letter has now become accessible, as a result of an auction of the "love letters."[6] In the following this passage will be set into the context of Einstein's early scientific interests and that of his attempt to obtain a doctorate in 1901. Although many aspects of this story have been discussed before, it is here recounted in its entirety in order to put together for the first time all available contemporary evidence showing the role of Einstein's interest in electron theory for the emergence of his approach to statistical mechanics.

3. Einstein's Early Controversies and his Failure to Obtain a Doctorate in 1901

3.1 A First Controversy: Einstein's Attempted Doctorate with H.F. Weber

In October 1900, Einstein was working on a doctorate in physics under the supervision of his ETH physics professor H.F. Weber (see the Questionnaire for Municipal Citizenship Applicants, 11–26 October 1900, Einstein 1987, Doc. 82, p. 270, and the editorial note in Einstein 1989, "Einstein's Dissertation on the Determination of Molecular Dimensions," pp. 170–182). The original choice of topic of the dissertation, which Einstein hoped to complete by Easter of 1901 (Mileva Marić to Helene Savić, 20 December 1900, Einstein 1987, Doc. 85, pp. 272–273), is not known. It may have been related to thermoelectricity, a subject on which several other students of Weber were doing research and on which Einstein himself had written his diploma thesis under Weber (see Einstein to Carl Seelig, 8 April 1952, and the editorial note in Einstein 1987, "Einstein on Thermal, Electrical, and Radiation Phenomena," pp. 235–237).

This first attempt at a doctorate, however, failed rather quickly. Einstein's correspondence with Marić testifies to a falling-out he had had with Weber by the beginning of spring 1901 (see, in particular, Einstein to Mileva Marić, 23 March 1901, Einstein 1987, Doc. 93, pp. 279–280, and Einstein to Mileva Marić, 27 March 1901, Einstein 1987, Doc. 94, pp. 281–283). As I have mentioned above, the background for this conflict is not known. A possible reason for it is that Weber expected his students to engage primarily in first-hand experimental work, whereas Einstein may have hoped that he could base theoretical conclusions on experimental work done by others.[7] In this case, Einstein's theoretical research for his dissertation may have been related not to thermoelectricity but to molecular forces in liquids, a subject on which he wrote his first paper, submitted in December 1900 (Einstein 1901). It is indeed difficult to imagine that Einstein's theoretical study of molecular forces as pursued in this paper could have been much to Weber's liking in view of the latter's preference for dissertations based on experimental work. But, in spite of the falling-out with Weber, Einstein continued to be interested in thermoelectricity as well and even developed a theory of his own on this subject (Mileva Marić to Einstein, second-half of May 1901, Einstein 1987, Doc. 110, p. 303).

The topic of molecular forces is mentioned as the subject of Einstein's doctoral work in a letter to his friend Marcel Grossmann in April 1901 (Einstein to Marcel Grossmann, 14 April 1901, Einstein 1987, Doc. 100, pp. 290–291). It is, however, likely that by that time Einstein was no longer working under Weber but under a new supervisor, Alfred Kleiner, professor at the University of Zurich. Einstein definitely attempted to obtain a doctorate under Kleiner in the fall of 1901, as I will discuss in detail below.

3.2 A Second Controversy: Einstein's Criticism of Drude and Boltzmann

Einstein's controversy with Drude took place in the middle of 1901, in the period of transition from his first supervisor Weber to Kleiner, and possibly after a shift of his dissertation topic in spring 1901 from thermoelectricity to molecular forces. Einstein nevertheless continued to follow the literature in thermoelectricity. He studied, in April 1901, Drude's electron theory of metals, finding it similar to his own approach to the subject (Einstein to Mileva Marić, 4 April 1901, Einstein 1987, Doc. 96, pp. 284–285). Drude's theory explains the empirically known connection between electric and thermal conduction, treating the assumed freely moving charged particles inside the metal on the basis of a specific, very simple gas model.[8] According to this model, the charged particles of each kind are assumed to move with the same speed within the metal; thermal conduction is assumed to be exclusively produced by collisions among the freely moving charged particles, while collisions with the atoms are assumed to play no role.

At the end of May, Einstein read with enthusiasm a paper by Max Reinganum who had shown that the explanation of the connection between electrical and thermal conduction does not in fact depend on such details of the atomistic model assumed (see Einstein to Mileva Marić, 28 May 1901, Einstein 1987, Doc. 111, p. 304, and Reinganum 1900). Reinganum rather argued that this explanation depends only on quite general statistical properties of the freely moving charge carriers, in particular on the validity of the equipartition theorem, that is, on the assumption that the mean kinetic energies of the charge carriers and of the atoms of the metal are equal. At the time he became familiar with Reinganum's work, Einstein decided not to publish his own theory but rather considered writing, between late May and early June, a personal letter to Drude (see Einstein to Mileva Marić, second-half of May 1901, Einstein 1987, Doc. 110, p. 303, and also Einstein to Mileva Marić, 4 June 1901, Einstein 1987, Doc. 112, p. 306). The reasons for Einstein's abandonment of his own theory have so far remained obscure; in section 4, I will argue that they are closely related to his reading of Reinganum.

As Einstein's letter to Drude is unknown, conclusions about its content have to rely exclusively on allusions to it in Einstein's other contemporary correspondence. What is clear from these allusions is only that he had formulated "two factual objections" to Drude's theory and that he had informed Drude that he was looking for a position (Einstein to Mileva Marić, 4 June 1901, Einstein 1987, Doc. 112, p. 306, and Einstein to Jost Winteler, 8 July 1901, Einstein 1987, Doc. 115, p. 310). Einstein received Drude's response in early July and forwarded it to Marić (Einstein to Mileva Marić, 7 July 1901, Einstein 1987, Doc. 114, p. 308). It was apparently very disappointing to him, probably both with regard to its scientific content and with regard to his hopes for a position. To Marić and to his fatherly friend Jost Winteler he announced that he would soon publish a polemic rejoinder to Drude (Einstein to Mileva Marić, 7 July 1901, Einstein 1987, Doc. 114, p. 308, and Einstein to Jost Winteler, 8 July 1901, Einstein 1987, Doc. 115, p. 310). This is virtually all that was known up to now about Drude's response, apart from the fact that, in his letter to Winteler, Einstein also referred to Drude's claim that an "infallible" colleague of his, whose name is not mentioned, was of the same opinion as Drude himself concerning the issue of Einstein's objections (Einstein to Jost Winteler, 8 July 1901, Einstein 1987, Doc. 115, p. 310).

The newly found passage stems from a letter by Marić written in early July, just after she had seen Drude's response.[9] Her letter to Einstein makes it clear how desperately they both must have waited for this response. She reacts with irony to the fact that that splendid fellow Drude has now finally sent a word. In the bohemian spirit of rebellion, which is documented by many letters of this period and which united the young couple against the rest of the world, and in particular against the scientific establishment, she mocked the court-like behavior of the masters of physics. Her remark about this behavior was occasioned by the way in which, in his response, Drude referred to another great master of classical physics, Ludwig Boltzmann. Drude's response must have been aimed, in Marić's sarcastic interpretation, at making it clear to a scientific novice such as Einstein that, of course, a great master like Boltzmann cannot have been wrong.

At first glance it is quite surprising that Drude should have mentioned Boltzmann in his response. It seems that Drude had felt the need to defend Boltzmann against Einstein's objections, rather than his own electron theory. But this single piece of additional information about Drude's response makes it now understandable why Einstein, immediately after this exchange, thoroughly reread Boltzmann's major work on gas theory and developed his own approach to statistical mechanics: in order to support his criticism of Drude which, in fact, must have been a criticism of Boltzmann as well. Einstein's subsequent scientific activities thus explain why Drude's reaction to Einstein's criticism contained a reference to Boltzmann.

Indeed, in early September 1901, Einstein wrote to his friend Marcel Grossmann that he had succeeded in filling a "gap" in Boltzmann's theory and that he was preparing a small publication about his findings (Einstein to Marcel Grossmann, 6 September 1901, Einstein 1987, Doc. 122, p. 315). The reference is most probably to an early draft of Einstein's first paper on statistical mechanics, which he eventually submitted in June 1902 (Einstein 1902b). Although the nature of the gap Einstein perceived is not clear from the letter to Grossmann, it can be reconstructed from this paper as well as from other documents concerning Einstein's statistical mechanics (see the editorial note in Einstein 1989, "Einstein on the Foundations of Statistical Physics," pp. 41–55). In his 1902 paper on the kinetic theory of thermal equilibrium, Einstein pointed to the achievements of the kinetic theory in the field of gas theory, but also to its failure to derive, from mechanics and probability theory alone, theorems on thermal equilibrium and the second law of thermodynamics for more general systems (Einstein 1902b, p. 417). He acknowledged that Maxwell and Boltzmann had come close to this goal, but also made it clear that this was the gap he intended to fill with his own paper. His paper suggests that he viewed Boltzmann's theory as lacking, in particular, a proof of the equipartition theorem for general physical systems in thermal equilibrium, that is, a proof that does not depend on the specific dynamic assumptions of the systems at hand.

The newly found passage makes it clear that Einstein's identification of this gap must have been closely related to his criticism of Drude's electron theory. From what we now know about Drude's response, it follows that at least one of Einstein's objections must have been directed at the statistical assumptions introduced by Drude and their justification by a kinetic theory of matter. This would indeed explain the reference to Boltzmann in Drude's reply. Drude apparently disputed the very existence of the gap that Einstein intended to fill not only in the statistical foundation of Drude's electron theory but also, more generally, in statistical physics as it had been developed by Boltzmann. After all, a master such as Boltzmann will surely have done it right, as Marić reported about Drude's response.

3.3 A Third Controversy: Einstein's Attempted Doctorate with Alfred Kleiner

My claim that Einstein's criticism of Drude was, at least in an essential part, directed against Boltzmann receives additional support from a reconstruction of Einstein's third controversy in the years 1901 and 1902. As mentioned above, he had turned, after his conflict with Weber, to Alfred Kleiner as the supervisor of his planned doctoral thesis. But although Einstein never had any personal falling-out with Kleiner as he did with Weber, this project also failed for reasons that have not been entirely clear. Here I will show that all available evidence points to the conclusion that Einstein included his new approach to statistical mechanics in his dissertation, along with applications of statistical physics to specific problems, and that the inclusion of this new approach may have been one of the reasons for Kleiner's advice to withdraw the dissertation.

In the period between mid-September and mid-November 1901, Einstein attempted to combine the various aspects of his research on the kinetic theory for a doctoral thesis (see Einstein to Swiss Patent Office, 18 December 1901, Einstein 1987, Doc. 129, p. 327; see also Mileva Marić to Helene Savić, 23 November – mid-December 1901, Einstein 1987, Doc. 125, p. 320). Did this thesis indeed comprise his work on statistical mechanics? This question can only be answered by a careful analysis of the rather indirect evidence that is available concerning this second early attempt by Einstein to obtain a doctorate. By mid-November he gave a draft of this thesis to Alfred Kleiner, along with an early draft paper dealing with the electrodynamics of moving bodies (Einstein to Mileva Marić, early November 1901, Einstein 1987, Doc. 123, p. 316, and Einstein to Mileva Marić, 13 November 1901, Doc. 124, p. 318). He submitted this dissertation officially to the University of Zurich on 23 November of that year (Receipt for the Return of Doctoral Fees, Einstein 1987, Doc. 132, p. 331). A contemporary letter by Marić describes the subject of the dissertation as a treatment of molecular forces in gases based on several known phenomena (Mileva Marić to Helene Savić, 23 November – mid-December 1901, Einstein 1987, Doc. 125, p. 320). In another contemporary letter Einstein discussed a problem he had discovered in applying his theory of molecular forces to liquids, which is another aspect of the kinetic theory possibly treated in his dissertation (see Einstein to Mileva Marić, 12 December 1901, Einstein 1987, Doc. 127, p. 324; see also Einstein to Grossmann, 14 April 1901, Einstein 1987, Doc. 100, pp. 290–291). From these references alone it is clear only that the dissertation was intended to cover a wide range of topics related to the kinetic theory but not whether it also included Einstein's work on statistical mechanics.

It is at this point that the puzzling remark in one of his contemporary letters, mentioned in section 2, adds an important piece to the picture, suggesting that statistical mechanics was indeed part of the dissertation. In mid-December Einstein wrote to Marić that he wondered what stance Drude would take once Kleiner accepts the dissertation (Einstein to Mileva Marić, 17 December 1901, Einstein 1987, Doc. 128, p. 326). This reference to Drude in the context of Einstein's dissertation is, at first glance, just as surprising as the remark in the newly-available passage mentioning Boltzmann in the context of Einstein's criticism of Drude's electron theory. It is difficult to account for this reference to Drude if the dissertation dealt only with molecular forces, as the direct references to its content seem to indicate. There are, in fact, no references to Einstein's dissertation (apart perhaps from this one) that point to a relation with thermoelectricity or electron theory, a relation that would, of course, explain Einstein's remark about the reaction he expected from Drude. But in light of the new evidence, Einstein's curiosity concerning this reaction does not need to be interpreted as implying that the dissertation also dealt with the latter's electron theory. It seems more likely that Einstein hoped to impress Drude by the way in which he had filled the gap in Boltzmann's approach to the kinetic theory, a gap whose very existence Drude apparently had denied.

In mid-December 1901, Kleiner had yet to read Einstein's dissertation (Einstein to Mileva Marić, 19 December 1901, Einstein 1987, Doc. 130, p. 328). The earliest available evidence for his reaction to it is dated February 1902 and represents a dramatic turn of events. On February 1, 1902, Einstein officially retracted his dissertation from the University of Zurich (Receipt for the Return of Doctoral Fees, Einstein 1987, Doc. 132, p. 331). A later biographer wrote, probably on the basis of a recollection by Einstein himself, that Kleiner had rejected the dissertation because of Einstein's sharp criticism of Ludwig Boltzmann (Kayser 1930, p. 69). This report fits well with the hypothesis that Einstein's dissertation combined his attempt to fill a gap in Boltzmann's approach to kinetic theory with a study of molecular forces in various applications, and—this cannot be excluded—perhaps even with allusions to the electron theory of metals and other generalizations of the kinetic theory. The hypothesis that the dissertation comprised an early version of Einstein's statistical mechanics receives support also from another contemporary letter in which he reports to Marić that a friend advised him to send to Boltzmann that part of his dissertation that refers to the latter's book, i.e., to Boltzmann's *Gastheorie*, which is the only work to which Einstein's 1902 paper on statistical mechanics makes reference (Boltzmann 1896 and 1898; Boltzmann 1898 is quoted in Einstein 1902b on pp. 420 and 427).

3.4 References to Einstein's Controversy with Drude in his Later Publications

In mid-December Einstein had written Mileva that, in case Kleiner rejected the dissertation, he would make that fact public, together with his work (Einstein to Mileva Marić, 17 December 1901, Einstein 1987, Doc. 128, p. 326). To Mileva and Jost Winteler he had earlier announced a polemic reply to Drude. The conflict with Weber, the controversy with Drude, and the failure of his dissertation project with Kleiner served to strengthen Einstein's perception that, in his early scientific endeavors, he was struggling against the established scientific authorities. It also confirmed and enhanced his aversion to authority in general (see the introduction to Renn and Schulmann 1992). To Winteler he wrote that he saw the blind belief in authority as the greatest enemy of truth. But Einstein also appears to have been somewhat discouraged by the reaction of the scientific establishment to his efforts. A direct rejoinder to Drude's negative response to his criticism could, in any case, not be identified among his publications. That no open attack on Drude exists is understandable. Apart from the fact that such an attack might have spoiled Einstein's ambitious plans for his career, it might not have been easy to publish it in view of the fact that Drude was himself the editor of the *Annalen der Physik* and a major leader of contemporary German physics (see the discussion in Jungnickel and McCormmach 1986, vol. 2, p. 309). But it is, of course, quite possible that at least some of Einstein's scientific arguments against Drude were not sacrified to such strategic considerations.

The newly found passage now makes it easier to scrutinize Einstein's publications of this period for traces of his failed dissertation and of his rejoinder to Drude. Some of his early papers apparently cover facets of this dissertation, first of all the paper on molecular forces in liquids (Einstein 1902a); then, according to the interpretation given here, at least the first of the three papers on statistical mechanics.[10] But neither of these papers contains any reference to Drude or to the electron theory of metals. The only explicit mention of Drude's electron theory in Einstein's early publications is found in his 1905 paper on the light quantum (Einstein 1905a, in Einstein 1989, Doc. 14, p.133). There he describes a physical system composed of gas particles, freely moving electrons, resonators, and radiation in thermal equilibrium, a system for which he assumes the validity of the equipartition theorem. In a footnote to the introduction of this assumption, Einstein refers to Drude's electron theory with the remark that this theory also depends on the assumption of the equipartition theorem. Although the footnote itself contains no criticism of Drude's theory, it is located in a section of Einstein's paper in which he discusses the catastrophic implications of an application of the equipartition theorem to classical radiation in thermal equilibrium, motivating the revolutionary proposal of the light quantum.

Implicitly, the context of this reference to Drude thereby makes it clear how highly problematic the application of the equipartition theorem to other systems than a gas could be, and hence also how much this application was in need of justification. It is ironical, and perhaps was meant to be so, that, instead of providing such a justification at this point (Einstein's papers on statistical physics are not quoted), he merely referred to the example of Drude's electron theory in order to justify his own application of the equipartition theorem to a system composed of gas particles, radiation, and electrons. It is indeed unlikely that Einstein was unaware of this crucial gap in his argument, which was the very part missing from Drude's theory.

The dissertation that Einstein finally submitted in 1905 also dealt with a facet of the kinetic theory, the determination of the dimensions of molecules in solution (Einstein 1905b). It treats the general question of establishing absolute molecular dimensions, but now in the context of a single, highly specialized problem;[11] it thus no longer bears direct evidence of Einstein's overarching ambitions in the years 1900 to 1902. The established authorities in physics, in this case represented by Alfred Kleiner, had probably made it impossible for Einstein to publish the diverse aspects of his exploration of the kinetic theory of matter under a single, unifying umbrella. The history of Einstein's first attempt to attain a doctorate in 1901 thus also explains the noteworthy splitting of his early publications into those dealing with the theoretical core and those dealing with the concrete implications of statistical mechanics.

In addition to splitting up the results of his 1901 dissertation into his various publications, Einstein may have sent privately to Boltzmann the part of the dissertation that was directly related to the Gastheorie, as his friend had advised him to do (see Einstein to Mileva Marić, 8 February 1902, Einstein 1987, Doc. 136, pp. 334–335). If so, his work on statistical mechanics failed to make any impression on Boltzmann, at least if its treatment in the last grand review of the kinetic theory Boltzmann wrote together with Nabl for Sommerfeld's *Encyclopädie* is taken as a gauge (Boltzmann and Nabl 1905, p. 549). There Einstein's first two papers on statistical mechanics are hidden, among several other papers by Boltzmann himself, in a footnote to a passage dealing with applications of the statistical method. Certainly Boltzmann and Nabl did not consider Einstein's contribution as filling an important gap in the kinetic theory, but rather saw it as one specialized contribution among many others. Therefore, the question remains as to how Einstein could have identified (or only believed he had identified) a gap that remained obscure to such masters of classical physics as Drude and Boltzmann. In order to attempt an answer to this question, we have to take another look not only at Einstein's perspective on the kinetic theory but also at the scientific context of this theory around the turn of the century.

4. The Interdisciplinary Potential of 19th Century Atomism and Einstein's Perspective

4.1 Atomism at the Turn of the Century: Growing Evidence and Growing Problems

Drude's electron theory and Boltzmann's kinetic theory of gases do not just happen to be two arbitrary subjects of interest to Einstein, but rather also share an important property with several other of his early research topics: they are two examples of the application of atomistic ideas to physical and chemical problems around the turn of the century.[12] Other prominent examples are Lorentz's atomistic version of Maxwell's electromagnetism, also often referred to as electron theory, the ion theory of electrolytic conduction, the kinetic theory of solutions, and the use of atomistic models in inorganic and organic chemistry. Whereas in antiquity and in early modern science atomism had served as a universal theory of nature, the scientific atomism of the late 19th century had become a specific and versatile conceptual tool in diversified branches of physics and chemistry.

Nevertheless, the different species of atomism clearly showed traces of their common ancestry. They all operated with the transposition of the concept of body, familiar from common experience, into an invisible microworld. Like bodies, atoms were thought to be distinguishable entities with a position in space and time, moveable independently of each other unless bound together. In addition they were ascribed other properties also familiar from the bodies of macroscopic experience, such as shape, rigidity, and electric charge. Which of these properties were ascribed to them in each single case depended on the specific context in which an atomistic microworld was introduced. In Lorentz's electromagnetism, for instance, the atomistic constituents of electricity were, of course, imagined to carry electric charge, whereas in the kinetic theory of gas, atoms were sometimes imagined as small elastic billiard balls. In chemistry, the complexity of the quantitative relations in chemical reactions was reduced to simple assumptions on the configuration of atoms in a molecule. In summary, in late 19th century physics and chemistry atomism, was widely considered to be a flexible working hypothesis useful for specific scientific explanations, while the question of the reality of atoms remained open to debate, in part because of the very heterogeneity evident in the ways in which atomistic hypotheses were deployed.

Scientists at the end of the 19th century were accordingly confronted with the problem of the compatibility of atomistic ideas elaborated in different contexts. In fact, as a consequence of the adaptation of the atomistic idea to the specific explanatory purposes of different branches of science, the atomism of the kinetic theory of gases was not obviously compatible with the atomism employed in theories of electromagnetism, let alone with that developed in chemistry. An example of such problems of compatibility is the clash between the chemical insights into the internal complexity of molecules and the failure of physical theories of atomism to account for this complexity. It was not only a matter of explaining the complex composition of molecules, but, more basically, of demonstrating that theories such as the kinetic theory of heat were not incompatible with the available chemical knowledge. It was in fact a well known problem for the kinetic theory that, in many cases, the thermal behavior of matter did not show evidence of this complex internal constitution. According to the equipartition theorem, every internal degree of freedom of an atomistic constituent of a gas should contribute equally to the mean energy and hence to the specific heat of the gas, thus implying the so-called Dulong-Petit law for specific heats, but empirical knowledge of specific heats suggested that this was not the case, given what was known about the internal degrees of freedom from the chemical composition of some gases and in particular from spectral analysis.[13]

In spite of such problems of compatibility between the different branches of atomism in the 19th century, problems which contributed to the skepticism with respect to the atomistic hypothesis, indications that it would continue to play a role in the conceptional foundations of physics became more numerous. Around the turn of the century, this growth consisted, on the one hand, in the sometimes surprising multiplication of ways in which Avogadro's number, a fundamental characteristic of the atomistic scale, could be established;[14] and, on the other hand, in novel opportunities for applications of the age-old atomistic idea. Naturally, such novel opportunities offered themselves, in particular, at the frontiers of experimental research, where new empirical knowledge was produced.[15] Examples are the new kinds of radiation for which attempts were made to decide whether they were waves or elementary particles: studies of the interaction between matter and radiation, such as the Zeeman and the photoelectric effects, which confirmed atomistic models of matter or gave rise to new ones: the study of solutions with the newly-developed ultramicroscope, which revealed several of them to be not homogeneous but suspensions of colloidal particles, etc. Measures of Avogadro's number could be gained from such different sources as the study of the capillarity of liquids, the kinetic theory of gases, experiments with thin layers, and, surprisingly, also from the theory of black body radiation. The agreement, at least in order of magnitude, between conclusions based on this diverse evidence about the size of the microworld not only increased confidence in the atomistic hypothesis but also made it seem ever more urgent to relate its different fields of application to each other and to establish a single atomistic model compatible with all of them.

4.2 Einstein's Atomism, his Perception of Contemporary Physics, and the Renouncement of his Theory of Thermoelectricity

The growing significance of atomism, and of the conceptual problems that came with its growth, was, of course, differently perceived by different contemporary scientists. Some scientists at the turn of the century preferred to deal only with the local applications of atomism relevant to their specific field of interest, disregarding the implications for the rest of physics and chemistry. For the young Einstein, on the other hand, atomism was a bond between different fields of contemporary science that allowed him not only to cherish the hope for conceptual unification (Einstein to Marcel Grossmann, 14 April 1901, Einstein 1987, Doc. 100, pp. 290–291), but also to perceive the conceptual clashes between different theories that to others appeared separated by disciplinary or subdisciplinary boundaries. Building up such an "interdisciplinary" vision of atomism, he could thus benefit from an unexploited potential of 19th century physics.

As I have pointed out elsewhere, he probably took up his studies of physics at the ETH with a bias in favor of atomism, a bias that was probably stimulated by his reading of popular scientific literature and that was strongly confirmed by his reading of Boltzmann (Renn 1993). In September 1900 he showed himself, in a letter to Marić, firmly convinced of Boltzmann's atomistic principles (Einstein to Mileva Marić, 13

September 1900, Einstein 1987, Doc. 75, pp. 260–261). Earlier he had already considered atomistic explanations of capillarity and of thermoelectricity (see the editorial notes in Einstein 1987, "Einstein on Thermal, Electrical, and Radiation Phenomena," pp. 235–237, and "Einstein on Molecular Forces," pp. 264–266). In the fall of 1900 he studied the theory of ions (Einstein to Mileva Marić, 3 October 1899, Einstein 1987, Doc. 79, p. 267); in early spring 1901 he developed a model of matter in which it is composed of atomistic electromagnetic resonators (see, e.g., Einstein to Mileva Marić, 23 March 1901, Einstein 1987, Doc. 93, pp. 279–280). In addition, he pondered in these years over an atomistic theory of light, performed radiation experiments, followed with enthusiasm the literature on the photoelectric effect and on electron theory, and worked on a theory of molecular forces applicable to gases as well as to liquids, comparing these forces to gravitation.[16]

Einstein also closely followed Planck's publications on black body radiation, one of them including a determination of Avogadro's number (see, e.g., Einstein to Mileva Marić, 10 April 1901, Einstein 1987, Doc. 97, pp. 286–287; see also Planck 1900). Just around the time when Einstein developed his criticism of Drude, in spring 1901, he raised doubts about Planck's justification for the equipartition of energy in his treatment of black body radiation (Einstein to Mileva Marić, 10 April 1901, Einstein 1987, Doc. 97, pp. 286–287). The fact that the equipartition theorem plays a key role both for the theory of heat radiation and for the electron theory of metals certainly contributed to placing it at the focus of Einstein's attention. As I have indicated, a derivation of this theorem from the most general principles possible was one of the problems he must have had in mind when he identified a "gap" in Boltzmann's *Gastheorie* later in 1901; how exactly he saw this gap and attempted to fill it will be discussed in the next section. But the broad range of Einstein's interests not only had an impact on his choice of problems to study, it also alerted him, as I have claimed above, to the "non-local" consequences of a model, i.e., to its implications beyond the narrow range of a specific application. The abandonment of his own theory of thermoelectricity may well be a case in point. In the following I will argue that this abandonment as well as Einstein's second "factual objection" to Drude were probably related to the latter's lack of concern with such implications.

The above mentioned problem of compatibility between the kinetic theory of heat and what was known about the behavior of matter from other contexts arises also for Drude's electron theory. In fact, Drude's explanation of electric and thermal conduction by an internal gas of charge carriers freely moving inside a metal is incompatible with the interpretation of the thermal behavior of metals by the classical kinetic theory. The freely moving charge carriers contribute many more degrees of freedom to the entire system than this interpretation admits, which rather suggests the model of a crystal-like rigid body with elastically bound atoms. This objection to the electron theory of metals was raised by several authors—but is not mentioned in Drude's papers—and was eventually resolved only in the context of quantum mechanics (see the discussion in Kaiser 1987). It seems not unlikely that it represents one of the reasons for Einstein's decision in 1901 not to follow up on his own ideas about thermoelectricity and to write a critical letter to Drude instead.

First of all, this hypothesis would solve a problem that was left open in the reconstruction presented in section 3. Indeed, Einstein's decision to abandon his own ideas about thermoelectricity cannot be related to the objection against Drude which we have discussed in that section and which was directed against the lack of justification for the application of the equipartition theorem in Drude's theory. On the contrary, when Einstein read, in late May 1901, Reinganum's paper on the central role of the equipartition theorem in the electron theory of metals, he showed himself completely convinced of its fundamental principles (Einstein to Mileva Marić, 28 May 1901, Einstein 1987, Doc. 111, p. 304). Why then should he have given up his pursuit of such a theory shortly after he became acquainted with this paper?[17]

The clue to Einstein's decision seems to lie in Reinganum's paper. In this paper Reinganum emphasizes the good agreement between Drude's theory and empirical knowledge and argues, as was mentioned earlier, that this agreement depends not so much on the details of Drude's atomistic model but on the assumption of the validity of the equipartition theorem lying behind this and similar models. But he also hints at a fundamental problem of an electron theory based on the hypothesis of free electrons, albeit only in a passing remark. Reinganum points to the incompatibility between Drude's electron theory of metals and a kinetic theory of specific heats that is based on Boltzmann's principles as a difficulty for either one of the two theories:

> Because of the experiments of Mr. Kaufmann on cathode rays, and because of Lorentz's theory, brilliantly confirmed by Zeeman's experiments, which ascribes to the bound electrons of luminiscent gases a number of degrees of freedom, the assumption of completely free electrons in metals does not appear too daring; nevertheless also the extended theory by Giese [which does not assume free electrons] merits consideration since, if one does assume free electrons, the theory of the Dulong-Petit law of specific heats, as it has been

> built up by Richarz according to Boltzmann's principles without so far leading to contradictions, would have to be replaced by a completely different theory of this law, at least if the number of electrons is comparable to that of the metal atoms. (Reinganum 1900, p. 401)

The way in which this passage connects Drude's theory to cathode ray experiments, to Zeeman's experiments and to Lorentz's electron theory could have only been to Einstein's liking. But once the significance of Reinganum's remark had dawned upon him, its upshot, the incompatibility of the assumption of free electrons in metals with the theory of specific heats following from the equipartition theorem, must have shattered his firm conviction in the principles of the electron theory of metals. The close temporal relation between Einstein's reading of Reinganum's paper, of which at first he noted only the aspects favorable to electron theory, and the decision to abandon his own approach, argues in favor of this incompatibility being the reason for his decision and also an important background for his controversy with Drude.[18]

There are, in addition, indications in the "love letters" that Einstein himself had been on the track of this problem in the atomistic explanation of thermoelectricity. In a letter to Marić of October 1899, one of the earliest letters in which he reported on his ponderings about the laws of thermoelectricity, he mentioned his intention to explore the contribution of charge carriers to the thermal behavior of a metal (Einstein to Mileva Marić, 10 October 1899, Einstein 1987, Doc. 58, p. 238). Pursuing this line of thought, Einstein may well have arrived at the conclusion that the contribution of charge carriers to specific heat was much too large to allow the explanation of thermoelectricity by an electron gas. His reading of Reinganum may have thus struck a familiar chord, particularly since in spring 1901 Einstein also attempted to explain deviations from the Dulong-Petit law of specific heats, i.e., from the implications of the equipartition theorem for the thermal behavior of solid bodies (Einstein to Mileva Marić, 23 March 1901, Einstein 1987, Doc. 93, pp. 279–280). In his famous 1907 publication about the explanation of such deviations on the basis of the quantum hypothesis, Einstein came back to the problematic consequences for the classical theory of specific heats of the assumption of mobile electrons inside a solid body and referred to a paper by Drude in support of this assumption.[19] In summary, the problem of specific heats was central to Einstein's thinking about the atomistic constitution of matter and is hence a likely reason for his growing skepticism towards an electron theory of metals.

It is plausible to surmise that Einstein's recognition of the problem of specific heats was related not only to the renouncement of his own theory of thermoelectricity but also to his second "factual" objection to Drude. Given the lack of direct contemporary evidence concerning Einstein's second objection to Drude, one can only rely on later references to the theory of electrons in metals and its problems in order to identify traces of this objection. There is, for instance, evidence that Einstein discussed the electron theory of metals with Paul Gruner who acknowledged Einstein's advice in a paper of 1909 (see Gruner 1909, p. 48; see also Paul Gruner to Einstein, 9 November 1908, Einstein 1993, Doc. 127, pp. 145–147). As the paper suggests, this advice pointed in a direction similar to Reinganum's and probably also to Einstein's observations on Drude's theory discussed in section 3, i.e., towards a generalization of the basic assumptions, and away from the specifics of a particular atomistic dynamics. Unfortunately, Gruner's paper offers no clue as to Einstein's objections against Drude.

In 1911, Einstein concerned himself again with the electron theory of metals, this time stimulated by conversations with Kamerlingh Onnes and Keesom concerning recent results on temperature dependence of electric conductivity, in particular, for low temperatures (Einstein to Hendrik A. Lorentz, 15 February 1911, Einstein 1993, Doc. 254, p. 281). Although this issue had only recently come to his attention, it does represent some general features relevant also for ordinary temperatures, features that might not have escaped Einstein's attention even in 1901 and that certainly continued to preoccupy him for years to come. In addition, the question of temperature dependence of electric conductivity is also closely related to the question of specific heat of metals since, according to the contemporary views, both conductivity and specific heat should depend on the density of free electrons. In fact, in an electron theory such as Drude's, electric conductivity is proportional to the density of free electrons and their mean free path. Experimentally it was known that electric conductivity increases with decreasing temperature, and the problem was how to explain this behavior in terms of temperature dependence of the density of free electrons, on the one hand, and in terms of temperature dependence of the mean free path, on the other hand. It was indeed difficult to disentangle the contribution of these two factors, as Einstein pointed out in a letter to Lorentz in early 1911 (ibid.). He also discussed this and related problems later that year in an exchange of letters with his friend Michele Besso (Einstein to Michele Besso, 11 September 1911, Einstein 1993, Doc. 283, p. 321; Einstein to Michele Besso, 21 October 1911, ibid., Doc. 296, p. 321; and Michele Besso to Einstein, 23 October 1911, ibid., Doc. 299, p. 342).

In a short paper of 1922, published in a *Gedenkboek* for Kamerlingh Onnes, Einstein succinctly summarized why temperature dependence represents a major problem for the formula for electric conductivity that results from Drude's electron theory of metals (Einstein 1922, pp. 430–432): Following this theory, let the specific resistance ω of a metal be given by:

$$\omega = \frac{2m}{\varepsilon^2} \frac{u}{nl},$$

where m is the mass, ε the charge of the electron, u its mean velocity, n the volume density of electrons and l their mean free path. The velocity u should be proportional to $\sqrt{T}$ according to the kinetic theory of heat, where T is the temperature. Einstein then argues in his 1922 paper that one should, at first sight, not expect l to depend significantly on temperature, while the number of electrically dissociated atoms, characterized by n, should rapidly grow with temperature. These assumptions are indeed plausible on the basis of a simple atomistic model of a metal such as that used by Drude as a conceptual background to his electron theory. But they are neither compatible with the Dulong-Petit law for specific heats nor with what was known about the temperature dependence of electric conductivity, since these assumptions imply that the resistance of metals rapidly *decreases* with temperature, contrary to its experimentally observed *increase* approximately in proportion to temperature (Seeliger 1921, p. 854). In his paper Einstein emphasizes that this problem constitutes a major defect of the electron theory of metals, even for the range of ordinary conduction phenomena, leaving aside superconductivity (Einstein 1922, p. 432). It is hence conceivable that this problem also constituted his second "factual" objection to Drude in 1901.

In 1925, Einstein returned to the electron theory of metals and once again discussed the problems of specific heat and of electrical conductivity (Einstein 1925, pp. 12–13). This paper makes it particularly clear that he considered failure of the electron theory of metals to provide an explanation for the missing contribution of the electrons to specific heat as a crucial weakness. But this time Einstein hoped to have found the reason for this lack. His explanation is based on his newly developed "quantum theory of the monatomic ideal gas," and an attempt to apply the concept of a saturated ideal gas to the electrons of a metal. Einstein believed that he could thus resolve the riddle of specific heat because his approach led to a considerably reduced density of the electrons that contribute to the thermal energy of a metal; but his very solution created a new problem for the explanation of electric conductivity. Since according to the electron theory of metals, electric conductivity depends on the product of electron density and mean free path, reduction of the electron density requires in fact the assumption of mean free paths larger than is physically reasonable. Later Einstein welcomed Sommerfeld's use of Fermi statistics in order to solve the riddle of specific heat, although Sommerfeld's approach suffered from a similar problem in explaining electric conductivity (see Einstein to Sommerfeld, 9 November 1927, published in Hermann 1968, pp. 111–112; and Kaiser 1987, pp. 294–295 for discussion).

A more thorough search for later references to the electron theory of metals in Einstein's writings might yield further hints, even about his early thinking on this subject. But my cursory examination of the evidence may be sufficient to illustrate how Einstein persisted in viewing the electron theory of metals within the framework of a larger atomistic picture of thermodynamic and electrodynamic phenomena. The search for coherency of this picture was also, it seems, the principal motivation for his early criticism of Drude.

5. Einstein's Statistical Mechanics and the Gap in Boltzmann's Theory

5.1 The *Gastheorie* from the Perspective of an Interdisciplinary Atomist

The wealth of contemporary applications of atomistics described in the previous section is hardly reflected in Boltzmann's *Gastheorie*, from which Einstein mainly drew his knowledge about the kinetic theory.[20] No wonder then that he observed, in a letter to Marić from April 1901, that in Boltzmann's *Gastheorie* he found too little emphasis on a comparison with reality (Einstein to Mileva Marić, 30 April 1901, Einstein 1987, Doc. 102, p. 294). Einstein's rich knowledge about the contemporary applications of atomism provided him with peculiar glasses through which he read this book, noticing shortcomings that easily escaped the attention of less broadly informed readers. But he was, on the other hand, less well prepared than other readers of the *Gastheorie*, in the sense that he was not familiar with Boltzmann's entire work.[21] He could therefore not perceive the book in the light of Boltzmann's earlier achievements and in that of the goals that had motivated the latter's research. Einstein rather brought his own interests to bear on the results presented in the *Gastheorie*, thus developing a new interpretation of these results by placing them into a new context.

In reading Boltzmann, Einstein must have naturally concentrated on the introduction of a statistical ensemble of systems, since such an ensemble is most suited for a generalization to systems other than gases.[22] Boltzmann's use of a virtual ensemble was intended to circumvent the practical impossibility of determining the time development of a system with many degrees of freedom. Analysis of its development in time is replaced by analysis of a statistical ensemble of copies of the system, all sharing the same dynamics but distributed over possible initial values compatible with given constraints on the system:

> The mathematically most perfect method would now consist in taking into account, for each state of a given warm body, the initial conditions, starting from which by chance the body, just in this case, reaches the thermal state that it now occupies unchanged for a long time. But since the same mean values result in any case, whatever the initial state might have been, they cannot be different from the mean values that we obtain when we conceive, instead of a single warm body, infinitely many of them that are completely independent of each other and that start, with equal heat content and equal external conditions, from all possible initial states. (Boltzmann 1898, § 35, pp. 102–103)

In his *Gastheorie* Boltzmann analyzed one particular kind of ensemble in great detail, which he called "Ergoden," and which is now called, following Gibbs, a "microcanonical ensemble":

> We now conceive again of an enormously great number of mechanical systems that all have the same constitution we have earlier described. The total energy E shall have again the same value. But for the rest, the coordinates and momenta of the various systems shall have, at the beginning of the time, the most diverse values. (Boltzmann 1898, § 32, p. 89)

For such a system Boltzmann demonstrated the equipartition of the "lebendige Kraft" (*vis viva*, i.e., kinetic energy) over the various "Momentoiden" (momentoids, a generalization of the concept of momentum to generalized coordinates; their number corresponds to that of the degrees of freedom of an atomistic constituent of the system) and concluded:

> Of course, this equality of the mean value of the *vis viva* corresponding to each momentoid is only proven for the presupposed (ergodic) distribution of states. This distribution of states is certainly a stationary one. There can be, however,

and in general there will be, other stationary distributions of states for which these theorems are not valid. (Boltzmann 1898, § 34, p. 101)

Einstein probably found it difficult to see how Boltzmann's derivation of the equipartition theorem from the assumption of such an ergodic or microcanonical ensemble could be put to practical use in treating radiation in thermal equilibrium or an electron gas inside a metal. In order to justify his application of the equipartition theorem to the electrons inside a metal, Drude had referred to a paper by Boltzmann on the theory of gas molecules (see Drude 1900a, p. 570; the paper by Boltzmann is Boltzmann 1868). Given how little was known about the dynamics of electrons inside a metal, this certainly was a problematic step. But the representation of the charge carriers in thermal equilibrium by an ergodic ensemble also could not have appealed to Einstein as an unproblematic alternative, since an ergodic ensemble presupposes a constant value of energy rather than of temperature as Drude had assumed. In the passage quoted above Boltzmann left it indeed open whether or not the equipartition of energy could be proven also for other stationary ensembles, e.g., an ensemble characterized by a constant value of the temperature.

Boltzmann's own admission of the difficulty of extending his results beyond gas theory may have confirmed Einstein's impression that there is indeed a gap in the *Gastheorie*. Boltzmann's attempt to apply his conclusions concerning the validity of the equipartition theorem also to solid and liquid bodies is essentially contained in the brief paragraph § 35, devoid of any formulas, and in a footnote (see Boltzmann 1898, § 35 and p. 126, note 1). The paragraph in question is introduced with the following cautious remark:

> Before I proceed to the application of the theorems presented until now to the theory of gases with polyatomic molecules, I will first add a completely general consideration which, however, does not rigorously rest on the mathematical standpoint but from the very beginning makes use of certain facts of experience. It nevertheless perhaps justifies the conjecture that the significance of these theorems is not restricted to the theory of polyatomic gas molecules. (Boltzmann 1898, p. 101)

The most important "fact of experience" that Boltzmann then uses as the starting point of his argument is that a body, whatever its initial state may have been, will appear to come to a stationary final state in which the observable mean values of its microscopic parameters always take on the same values. He continues with the justification of the usage of mean values taken over the statistical ensemble that I have quoted above:

> But since the same mean values result in any case, whatever the initial state might have been, they cannot be different from the mean values that we obtain when we conceive, instead of a single warm body, of infinitely many of them that are completely independent of each other and that start, with equal heat content and equal external conditions, from all possible initial states. (Boltzmann 1898, p. 103)

He then comes to a preliminary conclusion, again expressed in a very prudent tone:

> It therefore has a certain probability that the mean values found in § 34 are not only valid for the set of systems there defined but also for the stationary final state of any single warm body, [and] that, in particular, the equality of the mean *vis viva* corresponding to each momentoid [i.e., the validity of the equipartition theorem] is also in this case the condition for temperature equilibrium between different parts of a warm body. (Boltzmann 1898, p. 103)

In order to add plausibility to his argument, Boltzmann then briefly turns to the case of two physical systems in thermal equilibrium with each other. He justifies his assertion that the validity of the equipartition theorem is the characteristic property of their thermal equilibrium by considering a gas divided by a thermally conductive membrane, thus effectively making two systems out of one. His conclusion is that, for each molecule of one of the two gases, the mean kinetic energy of its center of gravity must be equal. Boltzmann finally extends his argument to the case in which one of the physical systems is an arbitrary body, by simply adding: "This mean *vis viva* should also be equal to the mean *vis viva* corresponding to an arbitrary momentoid, which determines the molecular motion of an arbitrary body in thermal equilibrium with the gas" (Boltzmann 1898, p. 104).

In a footnote to a later passage Boltzmann elaborates just a little bit more the idea of an arbitrary body in thermal equilibrium with a gas. He there justifies the extension of his theorems to such a body by considering it as representing a single gas molecule surrounded by the much larger mass of gas:

> We want to conceive a certain given solid or liquid [tropfbar flüssigen] body under the picture of an aggregation of n material points, which hence has $3n$ degrees of freedom, for instance the $3n$ orthogonal coordinates. If it is surrounded by a much larger mass of gas, it can, so to say, be considered as a

single gas molecule and the laws found in the text can be applied to it. (Boltzmann 1898, p. 126, note 1)

These extremely brief and isolated remarks do not, however, constitute a rigorous derivation of the validity of the equipartition theorem for general physical systems in thermal equilibrium, but rather mark even more vividly a weak spot in Boltzmann's exposition in the *Gastheorie*. The book, in particular, does not treat the canonical ensemble as a representation of a system in thermal equilibrium, let alone the equipartition of energy based on this representation. Since Einstein was quite aware of how precarious and central the validity of the equipartition theorem was in some of the applications of the kinetic theory of heat, in particular for the cases of electron theory and heat radiation, he must have identified this weak spot in Boltzmann's *Gastheorie* as a crucial gap. Boltzmann's line of argument, as I have sketched it, could have appeared to Einstein, on the other hand, almost like a blueprint of an approach to be worked out. In fact, Boltzmann's argument not only comprises essential ideas on which Einstein's statistical mechanics is based but also offers itself for a further mathematical elaboration because of its predominantly qualitative character. When Einstein took up this challenge, he did not know that such an elaboration had already been accomplished to a large extent by Boltzmann himself, albeit from a different perspective.

5.2 Statistical Mechanics from the Perspective of a Mechanical Atomist

It is impossible here exhaustively to review the numerous building blocks of statistical mechanics that can be traced back to various publications by Boltzmann.[23] Nevertheless, a few examples may suffice to indicate that the gap Einstein thought to have identified in Boltzmann was really no gap at all, at least as far as the crucial definitions and technical results taken by themselves are concerned. For example, Boltzmann had introduced the canonical ensemble in a paper of 1885, where he named it the "holode":[24]

> Let an arbitrary system be given whose state is characterized by the arbitrary coordinates $p_1, p_2, \ldots, p_g$; let the corresponding momenta be $r_1, r_2, \ldots, r_g$. We want to call them for short the coordinates p_g and the momenta r_g. Let the system be exposed to arbitrary internal and external forces; the former shall be conservative. Let ψ be the *vis viva* [i.e., the kinetic energy], χ the potential energy of the system.
>
> . . .

Case 1: We now conceive of very many N such systems, exactly equally constituted; each system completely independent of every other one. Let the number of all these systems for which the coordinates and momenta lie between the limits

p_1 and $p_1 + dp_1$, p_2 and $p_2 + dp_2$, . . . , r_g and $r_g + dr_g$

be:

$$dN = Ne^{-h(\chi+\psi)} \frac{\sqrt{\Delta} d\sigma d\tau}{\int\int e^{-h(\chi+\psi)} \sqrt{\Delta} d\sigma d\tau}$$

where

$$d\sigma = \Delta^{-1/2} dp_1 dp_2 \ldots dp_g, \; d\tau = dr_1 dr_2 \ldots dr_g.$$

(Boltzmann 1885, pp. 131–132)

Clearly, if one takes $h = 1/kT$ where k is Boltzmann's constant and T the absolute temperature, Boltzmann's expression corresponds to the modern definition of a canonical ensemble. In his paper, Boltzmann then determines the kinetic energy L and the potential energy Φ of the ensemble:

$$L = \frac{Ng}{2h}$$

and

$$\Phi = N \frac{\int \chi e^{-h\chi} d\sigma}{\int e^{-h\chi} d\sigma}.$$

He has thus established key properties of a canonical ensemble, and proceeds to show that it represents an equilibrium ensemble and that it works as a mechanical model for thermodynamics. Boltzmann's second example ("Case 2") is the ergodic or microcanonical ensemble, for which he demonstrates similar properties.

The principal aim of Boltzmann's introduction of a canonical ensemble (or of a "holode," as he called it) was not, however, to establish a statistical mechanics that would be applicable to an arbitrary physical system and suitable for deriving its thermodynamic properties, let alone for exploring its atomistic constitution. His intention was rather the opposite, solely to find a class of mechanical systems that show an analogy with the behavior of warm bodies. He thus elaborated a line of thought initiated by Helmholtz with the aim of studying the foundational question of the relation between mechanics and thermodynamics.[25] In the introduction to his paper Boltzmann developed this theme:[26]

> The most complete mechanical proof of the second law would obviously consist in showing that for each arbitrary mechanical process there are equations that are analogous to those of the theory of heat. But since this theorem does not, on the one hand, seem to be correct with this degree of generality; and since it is not possible, on the other hand, due to our ignorance of the essence of the so-called atoms, to exactly determine the mechanical conditions under which the thermal motion proceeds, the task emerges to study in which cases and to what extent the mechanical equations are analogous to those of the theory of heat. This cannot be a matter of postulating mechanical systems that are completely congruent to warm bodies but of finding all systems which more-or-less show an analogy with the behavior of warm bodies. The question was first posed in this manner by Mr. von Helmholtz, and I intend to pursue in the following the analogy he discovered between the systems which he designates as monocyclic and the theorems of the mechanical theory of heat in the case of some systems intimately related to the monocyclic ones. Before passing on to general theorems I will discuss some very special examples. (Boltzmann 1885, p. 122)

The special examples that Boltzmann discusses in the sequel merely serve to illustrate his theoretical program;[27] they deal with very special mechanical arrangements, such as a ring of continuous matter rotating around a central body, most of which are hardly directly relevant to problems of statistical physics.[28] Since Boltzmann was primarily concerned with foundational questions concerning the relation between mechanics and the theory of heat, it is not surprising that he mentioned the possibility of applying his approach to the treatment of solid and liquid bodies only in passing.[29] In the *Gastheorie* he did not even consider the canonical ensemble, particularly since its function for Boltzmann could just as well be served by an ergodic or microcanonical ensemble.[30]

This brings me to my second example of Boltzmann's hidden anticipation of results of statistical mechanics, this time taken directly from the *Gastheorie*. Although, in the *Gastheorie*, the only virtual ensemble considered is the ergodic one, Boltzmann did introduce there the mathematical expression for a canonical ensemble. But its physical interpretation did not refer to a virtual ensemble but rather to a real physical system—a gas—which is characterized by a probability distribution for its atomistic components. In his treatment of gases with polyatomic molecules,

Boltzmann began by assuming that, at a given initial time, the number of molecules of a certain kind whose center of gravity lies in a particular, infinitesimally small region of phase-space, is given by (Boltzmann 1898, § 37, p. 108):

$$A_1 e^{-2hE_1} dp_1 \ldots dq_\mu .$$

Here the position and state of a molecule (not of a member of a virtual ensemble!) are given by the generalized coordinates $p_1, p_2, \ldots, p_\mu$ and the generalized momenta $q_1, q_2, \ldots, q_\mu$ of its components; h is a constant depending on the temperature and A_1 a constant characteristic for the different species of molecules; finally, E_1 is the value of the sum of the total kinetic energy of a molecule and the potential energies due to intramolecular and external forces acting on the molecule at the given time.[31] That the above expression does not characterize a virtual ensemble is clear from the way it is introduced; it becomes entirely evident from the sequel of Boltzmann's argument. He considers in fact collisions between molecules in order to make plausible the stationary character of the assumed probability distribution. We have hence encountered a second case in which a mathematical expression given by Boltzmann is analytically the same as an expression in statistical mechanics but decidedly carries a different physical meaning.

A third example makes it clear that Boltzmann was not only familiar with the mathematical properties of the canonical ensemble but was also aware of its relation to the ergodic or microcanonical ensemble, even if again in a somewhat different manner from that later used in statistical mechanics. In a paper published in 1871 he considered an arbitrary body in thermal equilibrium with a gas (Boltzmann 1871b). In contrast with the extremely brief remarks in the *Gastheorie*, which we have discussed in the preceding subsection, he here offered a detailed quantitative analysis of the motion of the atoms of this body. He presupposed that the energy of the entire system is a constant of the motion and that the coordinates and velocities of its atomistic constituents take on all possible values compatible with this restriction. This system can thus be described by an ergodic or microcanonical ensemble, which is, however, not what he explicitly does in the section in question. For the sake of this particular argument Boltzmann instead focuses on the consideration of time averages. He assumes that the entire system consists of λ atoms, of which r belong to the body immersed in the mass of gas. In addition, both λ and r are taken to be large—with r, however, vanishingly small with respect to λ. The potential energy χ takes on the form:

$$\chi = \chi_1 + \chi_2$$

where χ_1 is a function of the r atoms of the body and χ_2 a function of the remaining $\lambda - r$ atoms of the gas.

Using a theorem of mechanics by Jacobi, Boltzmann first determines the average time during which the coordinates of the r atoms of the body lie between x_1 and $x_1 + dx_1$, etc., which he writes as:

$$dt = Ce^{-h\chi_1} dx_1 dy_1 \ldots dz_r$$

where h depends on the temperature as before and C is a constant. By taking into account also the velocities (or rather the momenta), an expression analogous to the probability distribution for a canonical ensemble can be derived by the same argument. In the same paper—albeit not in the same context—Boltzmann argues that time averages can, for systems such as the one considered here, be interchanged with ensemble averages (Boltzmann 1871b, pp. 277–278). If therefore the expression for the average time is reinterpreted as an expression concerning the number of systems in a virtual ensemble, it might be argued that Boltzmann had in effect demonstrated that a canonical ensemble represents a body in contact with a heat reservoir, which in turn can be represented by a microcanonical ensemble.

Such a reinterpretation of Boltzmann's result is in fact at the core of its discussion in the 1911 review paper by the Ehrenfests, where it is treated as an achievement with pioneering significance for statistical mechanics:

> It is actually the origin of the idea of representing the behavior of a body in thermal equilibrium by the average behavior of a canonical ensemble. In an ergodic system consisting of N molecules, let us consider a group of N' molecules, where N' may be a large number but still very small compared to N. Boltzmann obtains an expression for the relative length of time during which the state of these N' molecules lies in the region $dq_1^1 \ldots dp_r^{N'}$. This expression is
>
> (78) $dW = e^{-E'/\Theta} dq_1^1 \ldots dp_r^{N'}$,
>
> where E' is the total energy of the group of molecules in this state and $\Theta/2$ the time average of the kinetic energy per degree of freedom of the ergodic system. If we consider instead the corresponding stationarily distributed ergodic ensemble (Eq. 31a), the Eq. (78) will be proportional to the number of such individuals in the group for which the state of the N' molecules under consideration lies in the region $dq_1^1 \ldots dp_r^{N'}$. This is the form in which one finds the theorem in Maxwell's work (1878) [3]. Gibbs expresses it in the following

way (op. cit., p. 183): 'If a system of a great number of degrees of freedom is microcanonically distributed in phase, any very small part of it may be regarded as canonically distributed.' The part of the system of N' molecules in whose behavior we are interested is the body, while the whole ergodic system is this body together with a very large temperature bath. This is the way in which Einstein also uses the ergodic hypothesis and the microcanonical and canonical ensembles in two papers on the 'Kinetic Theory of Thermal Equilibrium and the Second Law of Thermodynamics' (1902, 1903). (Ehrenfest and Ehrenfest 1959/1990, p. 65)

The significance that the Ehrenfests attribute to Boltzmann's result, however, reflects then-existing knowledge of statistical mechanics. First of all, as we have seen, their interpretation of Boltzmann's expression in terms of a virtual ensemble, which is necessary in order to transform this result into a building block of statistical mechanics, was, although plausible, not actually carried out by Boltzmann himself. Secondly, his result played only a marginal role in his own work. It was conceived as merely supplementing the treatment of polyatomic gas molecules presented in an earlier paper (Boltzmann 1871a). Concerning the relationship between the two approaches, Boltzmann remarked in the concluding paragraph of his paper:

We thus arrive in a much easier way at what we have found there. Since, however, the demonstration that the hypothesis made in the present section is satisfied for warm bodies has not yet been given [i.e., that the coordinates and velocities of their atomistic constituents take on all possible values compatible with the energy equation], yes, that it is even possible [for this hypothesis] to be satisfied, I therefore have chosen in that treatise the path that is more complicated but free of any hypothesis. (Boltzmann 1871b, p. 287)

This attitude corresponds to that which Boltzmann also took in the *Gastheorie*. There he extensively discussed the other, more complicated approach to polyatomic gas molecules, which involves a detailed analysis of molecular dynamics and which I mentioned in the second example of this subsection. In the *Gastheorie*, he is merely alluding to the approach of his 1871 paper, in the context of the qualitative considerations of § 35, omitting the quantitative aspects discussed in the paper. By presenting the generalization to arbitrary bodies as being merely a "conjecture," Boltzmann, I argued above, provoked in at least one of his readers, namely Einstein, the impression that there was actually a gap in the *Gastheorie* (Boltzmann 1898, § 34, p. 101).

The few examples presented here illustrate that, although essential results of statistical mechanics were anticipated in the work of Boltzmann, they are there embedded in conceptual contexts different from that of statistical mechanics. Gibbs himself remarked that statistical mechanics brought a new perspective rather then entirely new technical details. In the preface to his 1902 *Elementary Principles of Statistical Mechanics*, he wrote, "The matter of the present volume consists in large measure of results which have been obtained by the investigators mentioned above [i.e., Maxwell and Boltzmann], although the point of view and the arrangement may be different" (Gibbs 1902, p. x), and in an earlier passage he noted:

> But although, as a matter of history, statistical mechanics owes its origin to investigations in thermodynamics, it seems eminently worthy of an independent development, both on account of the elegance and simplicity of its principles, and because it yields new results and places old truths in a new light in departments quite outside of thermodynamics. (Gibbs 1902, p. viii)

When Boltzmann, on the other hand, looked back in 1899 on his own research he felt that his motives and his perspective were different from those that, he felt, now prevailed:

> Each of these substances [caloric substance, electric and magnetic fluids, etc.] was conceived of as consisting of atoms, and the task of physics seemed confined for ever to ascertaining the law of action of the force acting at a distance between any two atoms and then to integrating the equations that followed from all these interactions under appropriate initial conditions.
>
> This was the stage of development of theoretical physics when I began my studies. How many things have changed since then! Indeed, when I look back on all these developments and revolutions I feel like a monument of ancient scientific memories. I would go further and say that I am the only one left who still grasped the old doctrines with unreserved enthusiasm—at any rate I am the only one who still fights for them as far as he can.
>
> . . .
>
> I therefore present myself to you as a reactionary, one who has stayed behind and remains enthusiastic for the old classical doctrines as against the men of today; but I do not believe that I am narrow-minded or blind to the advantages of the new doctrines. (Boltzmann 1974, p. 82)

Indeed, narrow-minded he was not; when Boltzmann commented in 1904 on the work of Gibbs he understood quite well that the latter's statistical mechanics, although based on a systematization of familiar results, nevertheless represents a novel approach with a scope much wider than that of "ascertaining the law of action of the force acting at a distance between any two atoms" and wider also than that of the kinetic theory of heat:

> The merit of having systematized this system, described in a sizable book and given a characteristic name, belongs to one of the greatest of American scientists, perhaps the greatest as regards pure abstract thought and theoretical research, namely Willard Gibbs, until his recent death professor at Yale College.
>
> . . .
>
> The wide perspective opening up, if we think of applying this science to the statistics of living beings, human society, sociology and so on, instead of only to mechanical bodies, can here only be hinted at in a few words. (Boltzmann 1974, p. 171)

The significance of a particular perspective for interpreting or reinterpreting physical results is hence not a matter of narrow- or open-mindedness but rather of the scientific context in which these results are immersed and which lends them a particular meaning. This scientific context was, as we have seen, different in Boltzmann's and in Einstein's case. In the following subsection I will reconstruct how Einstein succeeded, from his own perspective, in reinterpreting some of Boltzmann's results and in thereby laying, independently of Gibbs, the foundations of statistical mechanics. As Einstein's work on statistical mechanics has been extensively discussed in volume 2 of the *Collected Papers*, I can confine myself here to a few illustrative remarks.

5.3 Einstein's Statistical Mechanics as a Reinterpretation of Boltzmann's *Gastheorie*

Einstein certainly was in a more difficult position than Gibbs when he set out to fill the gap he perceived in Boltzmann. Whereas Gibbs was thoroughly familiar with Boltzmann's work, for example, with the crucial 1871 paper discussed above (see Gibbs 1902, p. viii), Einstein essentially knew only the *Gastheorie*. Apart from the qualitative considerations of § 35, his starting points for a quantitative treatment of the statistical properties of general mechanical systems were Boltzmann's introduction of the ensemble idea and the probability distribution for polyatomic molecules, introduced as the second example in the previous subsection. But Einstein's engagement with virtually all aspects of turn-of-the-century atomism not only enabled him to identify a gap in Boltzmann's argument, it also helped him to fill it. First of all, it would have been clear to Einstein that the validity of the probability distribution found in Boltzmann could not depend on features peculiar to gases, e.g., on the collisions between its molecules. In early 1901, Einstein had even speculated about an analogy between the energy distribution over the molecules of a gas and the energy distribution over the frequencies of radiation in thermal equilibrium.[32]

It was therefore possible for Einstein to reinterpret the probability distribution for the number of compound molecules in a gas with a given range of values for the coordinates and momenta of their components, quoted in the preceding subsection, as the probability distribution for the number of copies of a given physical system in a virtual ensemble under analogous conditions. In fact, Einstein's first paper on statistical mechanics contains the following expression characterizing a "canonical ensemble:"[33]

$$dN' = A'' e^{-2hE} dp_1 \ldots dq_n.$$

He interpreted the expression as giving the probability that the state variables of a system in thermal equilibrium with a system of infinitely large energy (the heat reservoir) lie in an infinitely small volume of phase space. The claim that Einstein first found the canonical ensemble by transferring the idea of an exponential distribution of the energies in a gas with polyatomic molecules from kinetic theory to the theory of statistical ensembles receives further support from the close similarity between Boltzmann's and Einstein's expressions for these two cases.

Although the reinterpretation of Boltzmann's probability distribution as defining a canonical ensemble disposed of the necessity of a detailed consideration of interactions between the atomistic constituents of a system, it did not, however, *per se* provide a physical justification for the assumption that this ensemble does in fact properly represent a physical system in thermal equilibrium. The approach by which Einstein attempts, in his first paper, to provide such a justification again closely follows Boltzmann, or rather, Einstein's reinterpretation of Boltzmann's arguments. In the *Gastheorie*, Boltzmann had argued, as we have seen, that his results could be extended to an arbitrary body by considering it as a single gas molecule in thermal equilibrium with a surrounding mass of gas. If now this relation, between the body as a large molecule and the surrounding gas, is transformed into a relation between ensembles, one arrives at Einstein's justification for the claim that the canonical ensemble represents a physical system in thermal equilibrium. In fact, the reinterpretation of Boltzmann's image suggests representing both the entire system and the body by such virtual ensembles, and leads accordingly to a study of their relation. Clearly the entire system could be represented by an ergodic or microcanonical ensemble and the body, at least tentatively, by a canonical ensemble. As the next and conclusive step, one had then to show, just as Gibbs put it, that "if a system of a great number of degrees of freedom is microcanonically distributed in phase, any very small part of it may be regarded as canonically distributed" (Gibbs 1902, p. 183).

Probably starting from Boltzmann's image, as I have suggested, Einstein thus transferred the idea of a small thermodynamic system in contact with a large heat reservoir, familiar from macroscopic thermodynamics, to the relation between the two kinds of ensembles. In his paper he indeed treated the canonical ensemble as a small thermodynamic system in contact with a large heat reservoir, which was in its turn represented by the microcanonical ensemble (see Einstein 1902b, in particular, § 3 and § 5). The complex line of argumentation that Einstein followed in his paper in order to derive the probability distribution of a system in thermal equilibrium is guided and rendered plausible, in spite of several shortcomings, by this fundamental idea.[34] His argument can be described as another successful transfer of properties familiar from the macroscopic world into a microworld, in this case one described by statistical ensembles.

This line of attack was well prepared by his earlier research experience: When Einstein approached the task of filling the gap in Boltzmann's work, he was already experienced in modifying, in various ways, the basic atomistic idea by additional assumptions that either transposed macroscopic properties into the microworld or that transferred properties assumed in one kind of atomistic theory to another one. An example for the first case is his (or rather Planck's) resonator model of matter, which makes the concept of a resonator, familiar from macroscopic electrodynamics, the constituent of an atomistic theory that promised to explain the interaction between matter and radiation (see Einstein to Mileva Marić, 23 March 1901, Einstein 1987, Doc. 93, pp. 279–280). An example of the second case is Einstein's (and, of course, Drude's) combination of the kinetic theory of matter with the atomistic theory of electricity in their ideas about electron theories of metal. By ascribing electrical charge to the atomistic constituents assumed in the kinetic theory of heat, these theories attempted, as discussed above, to explain the thermal properties of matter by the motion of these particles and its electrical properties by their charge. Einstein's reinterpretation of Boltzmann's results in his approach to statistical mechanics was thus facilitated by his experience with the transfer of properties from the macro- to the microworld.

It is remarkable that, when Boltzmann and Nabl briefly commented on Einstein's approach in their 1905 review, they identified Boltzmann's image of a body as a large molecule as its core idea. Their comment is implicit in a brief passage found in a section on the application of Liouville's theorem to the calculation of specific heats. The context of the passage is a discussion of the thermal equilibrium of polyatomic gases; in a footnote to the passage, Einstein's first two papers on statistical mechanics are cited along with several papers by Boltzmann:

> By application of the statistical method to arbitrary bodies (their treatment, so to say, as gas molecules with very many atoms) one can find mechanical systems which show full mechanical analogy with warm bodies,[35] not only a partial one as the cyclic systems of Helmholtz. (Boltzmann and Nabl 1905, p. 549)

In other words, Boltzmann and Nabl apparently considered Einstein's approach to be nothing but an elaboration of Boltzmann's image, as claimed above. In their brief comment they do not discuss, however, Einstein's reinterpretation of this image in terms of statistical ensembles. If at all, they allude to this aspect only in their phrase "application of the statistical method." What is more important, they present Einstein's approach in the same light in which Boltzmann had set his introduction of the canonical ensemble in his 1884 paper, which in fact is cited in Boltzmann's and Nabl's footnote: that is, as the elaboration of a mechanical analogy with warm bodies, in extension of Helmholtz's monocyclic systems. Nothing in their remark hints at a novel approach; Einstein's two papers rather appear as merely supplementing, at best, Boltzmann's own earlier publications, which in fact they do, at least from a technical point of view.[36] I take Boltzmann's and Nabl's reaction to Einstein's work as another confirmation of how much the creation of statistical mechanics was a matter of perspective.

6. Epilogue: the Ambivalent Success of Einstein's Atomism

Einstein had developed his statistical mechanics with the aim of covering a wider range of atomistic phenomena than those accessible to the kinetic theory of Maxwell and Boltzmann, and he wanted to contribute what he could to establish the existence of atoms beyond the possibility of doubt by the skeptics (Einstein 1979, p. 44). In this he was eminently successful, at least initially. Statistical mechanics helped him and others to augment the available evidence in favor of the atomistic constitution of matter and even provided hints at a discrete structure of radiation. Einstein's explanation of Brownian motion in 1905, together with its experimental confirmation by Perrin, was perhaps his most influential contribution in this regard, and it also was a consequence of his occupation with statistical mechanics.[37] Of course, not all of Einstein's youthful extrapolations of atomistic ideas had turned out to be as fortunate; his attempt at an electrodynamics of moving bodies based on a corpuscular theory of light, for instance, gave way to the introduction of a new framework of space and time that was essentially indifferent to the question of whether light was a wave or consisted of particles. But the accumulated evidence in favor of atomism that became available by the 1910s, and that was masterfully exposed in Perrin's comprehensive reviews, seemed to many to dispel doubts about the existence of atoms and thus represented a fulfillment of Einstein's original goals (see Nye 1972).

This evidence was taken, as described in section 4, from quite different physical and chemical, and otherwise unrelated contexts of research, and yet it all pointed to more or less the same value for Avogadro's number and other general features of a corpuscular structure of matter. Atomism was accordingly acquiring at this time an integrative, cross-disciplinary significance, linking large areas of physical and chemical knowledge to each other in new ways, and thus constituting for many a scientific reality for the material microworld independently of any single theory or discipline. As we have seen, atomism had played this role in Einstein's youthful endeavors, but it had now taken on this role also for the scientific community at large.

This wealth of evidence in favor of a discrete structure of matter augmented the challenge of a coherent description of this microworld. Since the properties of atoms were obtained by a transfer of properties from the macroworld into the microworld, there could be no *a priori* guarantee that the resulting combination of atomistic properties would yield a coherent picture. Atoms had to carry charge in order to account for electrical phenomena; they had to move rapidly in order to account for thermal phenomena; they had to have many internal degrees of freedom in order to be capable of explaining complex spectra, and yet not too many in order to give rise to the observed specific heats; they had to be able to combine in complicated ways into molecules in order to explain chemical composition, but they had to do all of this, if possible, on the basis of known physical laws (see Harman 1982, pp. 133–139 for an overview). In a word, the concept of the atom gradually came to occupy the position of the concept of the ether at the end of the 19th century, when it was overburdened with the tasks it had to fulfill and the properties it had to possess, in particular being an immoveable carrier of electromagnetic waves and yet showing no trace of the motion of masses passing through it.[38]

Einstein encountered the dilemma of finding a coherent picture of the microworld when he applied concepts of statistical physics to the problem of black body radiation in his 1905 light quantum paper. This application did disclose a surprising and novel aspect of the structure of radiation in thermal equilibrium, its particle-like behavior for low energy densities within the range of validity of Wien's radiation formula.[39] But Einstein's insight could not be brought into harmony with the ordinary way in which the microworld was furnished, following our macroscopic experience, that is, consists of either particles or states of a continuum.[40] With more success, Einstein employed statistical mechanics to unravel new aspects of the atomistic constitution of matter, in particular in his 1907 analysis of specific heats of solids (Einstein 1907). Eventually, however, it turned out that the same dilemma that Einstein had discovered in the theory of radiation also plagued the atomistic conception of matter. The history of quantum theory makes it clear that the elements that had often been used to populate the microworld, particles and states of a continuous medium, had difficulty in accounting for the growing wealth of empirical knowledge about the microstructure of matter. Statistical mechanics played a key role in the analysis of this knowledge and hence in the development of quantum theory. In this way, statistical mechanics was no longer an instrument for extracting from this empirical knowledge the evidence for a microworld made up of atoms as they were envisaged by the young Einstein, but rather became a tool for assembling insights into a new conceptual foundation of physics, beyond the classical dichotomy of particles and fields, and beyond any possibility of transferring properties from the macroworld to the microworld. Reminding the reader that Einstein never quite accepted this new foundation of physics, and that he had his problems in particular with its intrinsically statistical character, may provide a suitable but somewhat ironical ending to this reconstruction of the origins of statistical mechanics in the atomistic world picture of his youth.

Acknowledgments: I am particularly grateful to Johannes Fehr and David Gugerli, both ETH Zurich. Without their substantial help this paper simply could not have been written. With pleasure I also acknowledge the advice and the suggestions I received from Peter Damerow, based on a thorough reading of earlier versions of this paper. I thank Tilman Sauer and Gudrun Staedel-Schneider for a critical reading of the manuscript and Urs Schoepflin, as well as his colleagues from the library of the Max Planck Institute for the History of Science, for their unfailing help in providing me with literature. I also wish to thank Giovanni Gallavotti for first introducing me to Boltzmann's papers. Finally, I would like to warmly thank both Jed

Z. Buchwald and Silvan S. Schweber for thoroughly and critically reading the manuscript of this paper as well as for their substantial and stimulating advice.

NOTES

[1] For a history of the Maxwell-Boltzmann distribution, see Brush 1976, Chapter 10.

[2] This example is discussed in more detail in Renn 1993.

[3] In this paper I will not discuss the role of the ergodic hypothesis, which is usually at the focus of the history of statistical mechanics; for an excellent survey, see Brush 1976, Chapter 10; for a recent discussion of its role for Boltzmann, see Gallavotti 1994.

[4] The present treatment is focused on the equipartition theorem. For an extensive discussion of this as well as of other aspects of Einstein's approach to statistical mechanics, see Einstein 1989, the editorial note "Einstein on the Foundations of Statistical Physics," pp. 41–55.

[5] Einstein retracted the dissertation, probably on Kleiner's advice on 1 February 1902, see the Receipt for the Return of Doctoral Fees, Einstein 1987, Doc. 132, p. 331, and also the discussion below.

[6] I gratefully acknowledge the kindness of Mr. Felix de Marez Oyens, from Christie's, who pointed my attention to the missing page of the letter by Mileva Marić to Einstein, ca. 8 July 1901, Einstein 1987, Doc. 116. As, unfortunately, no copy of the page is available to me, my interpretation had to be based on a raw transcription of the passage in question.

[7] For further discussion, see the editorial note in Einstein 1989, "Einstein's Dissertation on the Determination of Molecular Dimensions," pp. 170–182.

[8] For an historical review of the electron theory of metals, see Kaiser 1987.

[9] For an incomplete publication of this letter, see Mileva Marić to Einstein, ca. 8 July 1901, Einstein 1987, Doc. 116, pp. 310–311. The following paraphrase is based on a raw transcription of the missing page.

[10] Einstein 1902b, 1903, and 1904. Einstein had announced his intention to publish his results on statistical mechanics in the *Annalen* as early as September 1901, see Einstein to Marcel Grossmann, 6 September 1901, Einstein 1987, Doc. 122, p. 315.

[11] This aspect is often overlooked; it is extensively discussed in the editorial note in Einstein 1989, "Einstein's Dissertation on the Determination of Molecular Dimensions," pp. 170–182.

[12] For a review, see Brush 1976, § 1.9; see also the editorial note in Einstein 1989, "Einstein on Brownian Motion," pp. 206–222, and the literature cited there.

[13] See Einstein 1907, pp. 184–185, and, for historical discussions, Brush 1976, § 10.8 and Harman 1982, pp. 134–139. The Dulong-Petit law states that the product of the atomic specific heat and atomic weight is—at least for a number of solid monatomic elements—constant.

[14] Avogadro's number is defined as the number of molecules per gram-mole; for an historical review of methods of its determination, see Brush 1976, pp. 75–78.

[15] For a brief review, see the editorial note in Einstein 1989, "Einstein's Dissertation on the Determination of Molecular Dimensions," pp. 170–173.

[16] See the discussion of these interests in Renn 1993. For Einstein's adherence to a corpuscular theory of light, see also the editorial note in Einstein 1989, "Einstein on the Theory of Relativity," p. 263.

[17] In his letter to Mileva Marić, second half of May 1901 (Einstein 1987, Doc. 110, pp. 303–304), Einstein mentioned his dissatisfaction with his theory of thermoelectricity and announced his intention to write to Drude. In his letter to Mileva Marić, 28 May 1901 (Einstein 1987, Doc. 111, p. 304), he mentioned the paper by Reinganum. Both letters carry no date but are dated on the basis of circumstantial evidence. Their sequence as given in Vol. 1 of the *Collected Papers* should be reversed and the correct date of Doc. 110 is probably 30 May 1901. This is suggested both by Einstein's treatment of the issue of a position in an insurance society and by his attitude to thermoelectricity in the two letters. Concerning Einstein's intention to search for a position in an insurance society, it is rather clear that Doc. 111 introduces the subject and Doc. 110 comes back to it. Concerning thermoelectricity it also appears more likely that Einstein's enthusiasm for the electron theory (Doc. 111) is followed by a disappointment with his own approach (Doc. 110) than vice versa.

[18] Drude later published a note of correction to his earlier papers where he did cite Reinganum (Drude 1902, p. 689), without, however, discussing this fundamental objection.

[19] See Einstein 1907, p. 185. Drude's paper (Drude 1904) deals with the interpretation of dispersion measurements, a topic also considered by Einstein as early as spring 1901, see Einstein to Mileva Marić, 27 March 1901, Einstein 1987, Doc. 94, p. 283.

[20] For evidence, see the recollection in Einstein 1979, p. 44, and, for a historical review, the editorial note in Einstein 1989, "Einstein on the Foundations of Statistical Physics," pp. 41–55.

[21] In his papers laying the foundations of statistical mechanics Einstein only quoted Boltzmann's *Gastheorie*; only in 1909 did he refer to an earlier work by Boltzmann: see the editorial note in Einstein 1989, "Einstein on the Foundations of Statistical Physics," in particular, p. 44.

[22] The ergodic ensemble introduced by Boltzmann later became the starting point for Einstein's paper on statistical mechanics, see Einstein 1902b, § 2.

[23] For such reviews, see Ehrenfest and Ehrenfest 1911 and Gallavotti 1994.

[24] See also Boltzmann 1872, pp. 401–402) and Boltzmann 1884, an earlier version of Boltzmann 1885.

[25] In a letter to Hermann von Helmholtz from 27 December 1884, Boltzmann writes (see Höflechner 1994, pp. II85–II 86):

> The introduction of the concept of monocyclic systems and the way in which you apply the Lagrangian equations also seems to me to be extraordinarily

important, and I myself believe to have refuted, in my most recent publication, Mr. Clausius who does not see in this [introduction of monocyclic systems] any essential progress with respect to my and Clausius' calculations. But one remark I nevertheless want to allow myself concerning this matter. If you apply your equations, gained from [considering] monocyclic systems, without much ado to gases, then this does not seem to me to be completely justified. The latter have a strong analogy with monocyclic systems, and both are apparently subordinated to a general concept which, if it is to comprise all monocyclic systems, would indeed have to be more generally defined than my concept of *Ergoden*. I do not, however, believe that the ideal gases of the kinetic gas theory are simply subordinated to the concept of monocyclic systems. Clausius mainly emphasizes the lack of a completely stationary motion in the former; however, this point may be remedied by means of the trick, used by me and Maxwell, [to introduce] the fiction of infinitely many congruent, independent systems. However, the condition of the constrained [gefesselten] monocyclic systems that each gas molecule should be capable of performing by itself exactly the same motion which it performs in the gas, and that the connection of the different molecules in the gas consists in nothing else but a congruent regulation of their velocities, this condition does not seem to me to be fulfilled.

[26] For a discussion of Helmholtz's approach and its relation to that of Boltzmann, see Klein 1972.

[27] For a discussion of Boltzmann's approach and its relation to that of Helmholtz, see Klein 1972 and 1974, as well as the detailed studies Bierhalter 1981, 1987, and 1993.

[28] Consider, e.g., the first example in Boltzmann 1885, p. 123:

> Let a mass point move according to Newton's law of gravitation around a fixed central body *0* in an elliptic trajectory. Obviously, this motion is not monocyclic but we can transform it into a monocyclic one by using a trick which I first used in the first paragraph of my treatise "Some General Theorems on Thermal Equilibrium" and which was further analyzed afterwards by Maxwell. We imagine the entire elliptic trajectory covered with mass whose density at each point (i.e., the mass contained in one unit length of the trajectory) is supposed to be such that, while in the course of time mass is continuously flowing through each cross-section of the trajectory, the density in each point of the trajectory remains unchanged. If a ring of Saturn would consist of a homogeneous liquid or a homogeneous swarm of solid particles, it could provide an example for the motion under consideration if the cross-section of the ring is chosen appropriately. External forces may accelerate the motion or change the eccentricity of the trajectory; but the trajectory may also be changed by a very slow increase or diminution of the mass of the central body, while external work is thus transferred to the moveable ring, since it generally does not exert the same attraction on the mass that is added in the case of an increase of the mass of the central body as it does on the mass to be

taken away in the case of diminution.

[29] This passing reference is found in a passage in which Boltzmann takes issue with Helmholtz, pointing out that he had overcome some of the weaknesses of the latter's original treatment of monocyclic systems (see also note 25 above for Boltzmann's view of Helmholtz's approach):

> My equations, on the other hand, are not restricted to the case that the derivatives of all rapidly variable parameters with respect to time are functions of a single variable. On the contrary, in my equations arbitrarily many independent derivatives q_g of rapidly variable parameters p_g may occur. The q_g themselves may again be rapidly variable; only the mean *vis viva* [i.e., the kinetic energy] is determined by a single variable, the temperature. Thus, the ideal gases and those classes of *Monoden*, for which already Maxwell l.c. has demonstrated that they probably correspond to the solid and liquid [tropfbar] bodies occurring in nature, are contained as special cases in my formulae. (Boltzmann 1885, p. 142)

For critical discussions of Helmholtz's approach, see Bierhalter 1987 and 1993; for an assessment of Boltzmann's contribution to this line of research, see Bierhalter 1993.

[30] It was hence not any "obscurity" that was responsible for the neglect of this particular contribution of Boltzmann's, as Gallavotti suggests (Gallavotti 1994, p. 1572), but his different outlook on what today are problems of statistical mechanics. Obscurity in hindsight is often a side-effect of historical development, see the discussion in Damerow et al. 1992, p. 5.

[31] The analogy with the Maxwell-Boltzmann distribution for monatomic gases is discussed in Boltzmann 1898, pp. 121–122.

[32] See Einstein to Mileva Marić, 30 April 1901, Einstein 1987, Doc. 102, pp. 294–295. Drude did not take into account such an energy distribution for the free electrons inside a metal; indeed, its derivation from the dynamics of the electron motion inside the metal would have been purely speculative.

[33] Einstein 1902b, p. 422. Einstein did not use this terminology.

[34] See, e.g., Einstein's exchange with Paul Hertz on certain difficulties of Einstein's argumentation, which is briefly summarized in note 20 to Einstein 1902b, Einstein 1989, p. 74.

[35] The footnote to this passage contains the following references: Boltzmann 1871a, 1877, 1884, 1887; Einstein 1902b, 1903. Boltzmann 1885, discussed above, is an extended version of Boltzmann 1884.

[36] The similarity of Einstein's formalism with that of Boltzmann, together with the fact that Einstein designated his own approach as just filling a gap, may account for the remarkable difference between Boltzmann's reaction to Gibbs, on the one hand, and to Einstein, on the other. The greater prominence of Gibbs may, of course, also have played a role.

[37] See the editorial note in Einstein 1989, "Einstein on Brownian Motion," pp. 206–222 and the literature cited there.

[38] See Einstein 1920 for a discussion of the ether as a concept overloaded with requirements.

[39] See Einstein 1905a. The dependence of Einstein's paper on the light quantum on his earlier work on statistical physics is analyzed in Klein 1963.

[40] For a reconstruction of this conflict and its empirical roots, see Wheaton 1983.

References

Bierhalter, Günter (1981). "Zu Hermann von Helmholtzens mechanischer Grundlegung der Wärmelehre aus dem Jahre 1884." *Archive for the History of Exact Sciences* 25: 71–84.

— (1987). "Wie erfolgreich waren die im 19. Jahrhundert betriebenen Versuche einer mechanischen Grundlegung des zweiten Hauptsatzes der Thermodynamik?" *Archive for the History of Exact Sciences* 37: 77–99.

— (1993). "Helmholtz's Mechanical Foundation of Thermodynamics." In *Hermann von Helmholtz and the Foundations of Nineteenth-Century Science.* David Cahan, ed. Berkeley: University of California Press, pp. 432–458.

Boltzmann, Ludwig (1868). "Studien über das Gleichgewicht der lebendigen Kraft zwischen bewegten materiellen Punkten."*Kaiserliche Akademie der Wissenschaften. Mathematisch-naturwissenschaftlichen Classe. Sitzungsberichte* 58: 517–560. Reprinted Boltzmann 1968, Vol. 1, pp. 49–96.

— (1871a). "Über das Wärmegleichgewicht zwischen mehratomigen Gasmolekülen." *Kaiserliche Akademie der Wissenschaften. Mathematisch-naturwissenschaftlichen Classe. Sitzungsberichte* 63: 397–418. Reprinted in Boltzmann 1968,Vol. 1, pp. 237–258.

— (1871b). "Einige allgemeine Sätze über Wärmegleichgewicht." *Kaiserliche Akademie der Wissenschaften. Mathematisch-naturwissenschaftlichen Classe. Sitzungsberichte* 63: 679–711. Reprinted in Boltzmann 1968, Vol. 1, pp. 259–287.

— (1872). "Weitere Studien über das Wärmegleichgewicht unter Gasmolekülen." *Kaiserliche Akademie der Wissenschaften. Mathematisch-naturwissenschaftlichen Classe. Sitzungsberichte* 66: 275–370. Reprinted in Boltzmann 1968, Vol. 1, pp. 316–402.

— (1877). "Über die Beziehung zwischen dem zweiten Hauptsatze der mechanischen Wärmetheorie und der Wahrscheinlichkeitsrechnung respektive den Sätzen über das Wärmegleichgewicht." *Kaiserliche Akademie der Wissenschaften. Mathematisch-naturwissenschaftlichen Classe. Sitzungsberichte* 76: 373–435. Reprinted in Boltzmann 1968, Vol. 2, pp. 164–223.

—(1884). "Über die Eigenschaften monocyclischer und anderer damit verwandter Systeme." *Kaiserliche Akademie der Wissenschaften. Mathematisch-naturwissenschaftlichen Classe. Sitzungsberichte* 90: 231–245.

—(1885). "Über die Eigenschaften monocyclischer und anderer damit verwandter Systeme." *Journal für die reine und angewandte Mathematik* 98: 68–98. Reprinted in Boltzmann 1968, Vol. 3, pp. 122–152.

— (1887). "Ueber die mechanischen Analogien des zweiten Hauptsatzes der Thermodynamic." *Journal für die reine und angewandte Mathematik* 100: 201–212 (text submitted 1885). Reprinted in Boltzmann 1968, Vol. 3, pp. 258–271.

— (1896). *Vorlesungen über Gastheorie.* Part 1. Leipzig: Johann Ambrosius Barth.

— (1898). *Vorlesungen über Gastheorie.* Part 2. Leipzig: Johann Ambrosius Barth.

—(1968). Wissenschaftliche Abhandlungen, 3 vols. F. Hasenöhrl, ed. New York: Chelsea.

— (1974). *Theoretical Physics and Philosophical Problems. Selected Writings, Vienna Circle Collection.* Dordrecht and Boston: Reidel.

Boltzmann, Ludwig, and Nabl, Josef (1905). "Kinetische Theorie der Materie." In *Encyclopädie der mathematischen Wissenschaft. Mit Einschluss ihrer Anwendungen. Bd. 5: Physik, Teil 1.* A. Sommerfeld, ed. Leipzig: B. G. Teubner, 1904–1922, pp. 493–557. (Text submitted 1905.)

Brush, Steven (1976). *The Kind of Motion We Call Heat: A History of Kinetic Theory of Gases in the 19th Century. Book 1—Physics and the Atomists, Book 2—Statistical Physics and Irreversible Processes.* Amsterdam: North-Holland.

Damerow, Peter; Freudenthal, Gideon; McLaughlin, Peter; and Renn, Jürgen (1992). *Exploring the Limits of Preclassical Mechanics. A Study of Conceptual Development in Early Modern Science: Free Fall and Compounded Motion in the Work of Descartes, Galileo, and Beeckman.* New York: Springer-Verlag.

Drude, Paul (1900a). "Zur Elektronentheorie der Metalle: 1. Teil." *Annalen der Physik* 3 oder 1: 566–613.

— (1900b). "Zur Elektronentheorie der Metalle: 2. Teil." *Annalen der Physik* 4 oder 3, no. 11: 369–402.

— (1902). "Zur Elektronentheorie der Metalle." *Annalen der Physik* 7: 687–692.

— (1904). "Optische Eigenschaften und Elektronentheorie. I. Teil." *Annalen der Physik* 14: 677–725.

Ehrenfest, Paul and Ehrenfest, Tatiana (1911). "Begriffliche Grundlagen der statistischen Auffassung in der Mechanik." In *Die Encyclopädie der mathematischen Wissenschaften. Mit Einschluss ihrer Anwendungen. Bd. 4: Mechanik, Teil 2, Nr. 6.* F. Klein and C. Müller, eds. Leipzig: B. G. Teubner, 1904–1922, pp. 3–90. (Text submitted 1911.)

— (1959/1990): *The Conceptual Foundations of the Statistical Approach in Mechanics*. Ithaca: The Cornell University Press; reprint New York: Dover, 1990. (Transl. of Ehrenfest and Ehrenfest 1911.)

Einstein, Albert (1901). "Folgerungen aus den Capillaritätserscheinungen." *Annalen der Physik* 4: 513–52; reprinted in Einstein 1989, Doc. 1, pp. 10–20.

— (1902a). "Über die thermodynamische Theorie der Potentialdifferenz zwischen Metallen und vollständig dissociirten Lösungen ihrer Salze, und über eine elektrische Methode zur Erforschung der Molekularkräfte." *Annalen der Physik* 8: 798–814; reprinted in Einstein 1989, Doc. 2, pp. 23–39.

— (1902b). "Kinetische Theorie des Wärmegleichgewichtes und des zweiten Hauptsatzes der Thermodynamik." *Annalen der Physik* 9 (1902): 417–433; reprinted in Einstein 1989, Doc. 3, pp. 57–73.

— (1903). "Eine Theorie der Grundlagen der Thermodynamik." *Annalen der Physik* 11: 170–187; reprinted in Einstein 1989, Doc. 4, pp. 77–94.

— (1904). "Zur allgemeinen molekularen Theorie der Wärme." *Annalen der Physik* 14: 354–62; reprinted in Einstein 1989, Doc. 5, pp. 99–107.

— (1905a). "Über einen die Erzeugung und Verwandlung des Lichtes betreffenden heuristischen Gesichtspunkt." *Annalen der Physik* 1: 132–148; reprinted in Einstein 1989, Doc. 14, pp. 150–166.

— (1905b). *Eine neue Bestimmung der Moleküldimensionen. Inaugural-Dissertation zur Erlangung der philosophischen Doktorwürde der hohen Philosophischen Fakultät (Mathematisch-Naturwissenschaftliche Sektion) der Universität Zürich, vorgelegt 1905*. Bern: Buchdruckerei K.J. Wyss, 1906; reprinted in Einstein 1989, Doc. 15, pp. 186–202.

— (1905c). "Über die von der molekularkinetischen Theorie der Wärme geförderte Bewegung von in ruhenden Flüssigkeiten suspendierten Teilchen." *Annalen der Physik* 17: 549–56; reprinted in Einstein 1989, Doc. 16, pp. 224–235.

— (1907). "Die Planck'sche Theorie der Strahlung und die Theorie der spezifischen Wärme." *Annalen der Physik* 2: 180–190; reprinted in Einstein 1989, Doc. 38, pp. 379–389.

— (1920). *Äther und Relativitätstheorie. Rede. Gehalten am 5. Mai 1920 an der Reichs-Universität zu Leiden*. Berlin: Julius Springer.

— (1922). "Theoretische Bemerkungen zur Supraleitung der Metalle." In *Het Natuurkundig Laboratorium der Rijksuniversiteit te Leiden in de Jaren 1904–1922. Gedenkboek aangeboden aan H. Kamerlingh Onnes*. Leiden: E. Ijdo, pp. 429–435.

— (1925). "Quantentheorie des einatomigen idealen Gases. Zweite Abhandlung." *Preussische Akademie der Wissenschaften* (Berlin) *Sitzungsberichte*: 3–14.

— (1979). *Autobiographical Notes. A Centennial Edition*. P.A. Schilpp, ed. La Salle, Ill.: Open Court, 1979.

— (1987). *The Collected Papers of Albert Einstein*. Vol. 1, *The Early Years, 1879–1902*. John Stachel, et al., eds. Princeton: Princeton University Press.

— (1989). *The Collected Papers of Albert Einstein.* Vol. 2, *The Swiss Years: Writings, 1900–1909.* John Stachel, et al., eds. Princeton: Princeton University Press.

— (1993). *The Collected Papers of Albert Einstein.* Vol. 5, *The Swiss Years: Correspondence, 1902–1914.* Martin J. Klein, et al., eds. Princeton: Princeton University Press.

Gallavotti, Giovanni (1994). "Ergodicity, Ensembles, Irreversibility in Boltzmann and Beyond." *Journal of Statistical Physics* 78: 1571–1589.

Gibbs, Josiah Williard (1902). *Elementary Principles in Statistical Mechanics Developed with Especial Reference to the Rational Foundation of Thermodynamics, Yale Bicentennial Publications.* New York: Charles Scribner's Sons.

Gruner, Paul (1909). "Über die Bewegung der freien Elektronen in den Metallen." *Physikalische Zeitschrift* 10: 48–51.

Harman, Peter Michael (1982). *Energy, Force, and Matter. The Conceptual Development of Nineteenth-Century Physics.* Cambridge, UK: Cambridge University Press.

Hermann, Armin, ed. (1968). *Albert Einstein/Arnold Sommerfeld. Briefwechsel.* Basel and Stuttgart: Schwabe.

Höflechner, Werner, ed. (1994). *Ludwig Boltzmann. Leben und Briefe.* Graz: Akademische Druck- u. Verlagsanstalt.

Jungnickel, Christa, and McCormmach, Russel (1986). *Intellectual Mastery of Nature: Theoretical Physics from Ohm to Einstein. Vol 2: The Now Mighty Theoretical Physics 1870–1925.* Chicago: The University of Chicago Press.

Kaiser, Walter (1987). "Early theories of the electron gas." *Historical Studies in the Physical and Biological Sciences* 17: 271–297.

Kayser, Rudolf [Anton Reiser, pseud.] (1930). *Albert Einstein: A Biographical Portrait.* New York: Albert and Charles Boni.

Klein, Martin J. (1963). "Einstein's First Paper on Quanta." *The Natural Philosopher* 2: 59–86.

— (1970). *Paul Ehrenfest. Vol. 1: The Making of a Theoretical Physicist.* Amsterdam: North-Holland/New York: American Elsevier.

— (1972). "Mechanical Explanation at the End of the Nineteenth Century." *Centaurus* 17: 58–81.

— (1974). "Boltzmann, Monocycles, and Mechanical Explanation." In *Philosophical Foundations of Science.* Boston Studies in the Philosophy of Science, vol. 11. Raymond J. Seeger and Robert S. Cohen, eds. Dordrecht, Boston: D. Reidel, pp. 155–75.

Nye, Mary Jo (1972). *Molecular Reality: A Perspective on the Scientific Work of Jean Perrin.* London: MacDonald.

Planck, Max (1900). "Zur Theorie des Gesetzes der Energieverteilung im Normalspectrum." *Deutsche Physikalische Gesellschaft. Verhandlungen* 2: 237–245.

Reinganum, Maximilian (1900). "Theoretische Bestimmung des Verhältnisses von Wärme und Elektricitätsleitung der Metalle aus der Drude'schen Elektronentheorie." *Annalen der Physik* 2: 398–403.

Renn, Jürgen (1993). "Einstein as a Disciple of Galileo: A Comparative Study of Conceptual Development in Physics." *Science in Context* 6: 311–341.

Renn, Jürgen, and Schulmann, Robert, eds. (1992). *Albert Einstein—Mileva Marić. The Love Letters*. Princeton: Princeton University Press.

Seeliger, Rudolf (1921). "Elektronentheorie der Metalle." In *Encyklopädie der mathematischen Wissenschaften. Mit Einschluss ihrer Anwendungen. Bd. 5: Physik, Teil 2.* A. Sommerfeld, ed. Leipzig: B.G. Teubner 1904–1922, pp. 777–878. (Text submitted 1921.)

Wheaton, Bruce R. (1983). *The Tiger and the Shark: Empirical Roots of Wave-Particle Dualism*. Cambridge, UK: Cambridge University Press.

The Construction of the Special Theory: Some Queries and Considerations

Robert Rynasiewicz

1. Introduction

To the scientifically literate and the public at large, Einstein is synonymous with genius, and nothing fits the image of genius more than immediate and effortless insight. When asked by the biographer Carl Seelig if a definite birth date could be assigned to the theory of relativity, Einstein wrote back: "Between the conception of the idea of the special theory of relativity and the completion of the corresponding published paper there passed five or six weeks" (Seelig 1960, p. 114). This is the stuff of which legends are made—a revolution virtually overnight. But history is often more complicated than we might care to think. Einstein immediately qualified his reply: "However, it would hardly be correct to call this a birth date, in that really the arguments and building stones had been prepared for many years prior to this, although without precipitating the ultimate decision" (Seelig 1960, p. 114).

This need not surprise us. It took eight long years, beginning in 1907 with the seminal insight into the equivalence of a uniformly accelerating frame with a uniform gravitational field, for Einstein to arrive finally, in late 1915, at the field equations of general relativity. This development, involving a number of false starts and blind alleys, is largely documented by a sequence of published articles and in further detail by a wealth of contemporary correspondence and manuscript material.

By contrast, Einstein published nothing on the topic of the optics and electrodynamics of moving bodies prior to 1905, although we know for a fact that he had been concerned with this topic for at least seven years. There are no surviving manuscripts or notebooks, and only a scattering of clues from his correspondence during this period, even taking into account some forty recently discovered letters to Mileva Marić from the years 1898 to 1901. Beyond this lies a fragmentary body of later reminiscences and

Einstein Studies, vol. 8: Einstein: The Formative Years, pp. 159–201.

remarks, some of it reaching us only indirectly through the reports of others.

Barring the unforeseen discovery of new documents, we cannot hope to establish the path followed by Einstein in the construction of the special theory with anywhere near the degree of detail we can for the general theory. It may seem that there is nothing left to do but to reaffirm our ignorance. Nevertheless it is virtually impossible to ponder this momentous episode in the history of thought without also imagining the few fixed points embedded in some overall scenario that renders them intelligible. What scenario we imagine is inevitably governed by tacit assumptions and expectations. My purpose in what follows is to formulate a number of questions and considerations that, depending on how they are answered, will largely shape the picture that emerges. Some of these may prove ultimately unanswerable. Even so, it is crucial to draw attention to presuppositions that might otherwise be overlooked.

2. The Relativity of Simultaneity

The spring of 1905 was truly a busy one for Einstein. In apparently short order he completed four memorable works:

17 March: "On a Heuristic Point of View Concerning the Production and Transformation of Light." (Einstein 1905a; dated Bern, 17 March 1905; received on 18 March 1905 by the *Annalen der Physik.*)

30 April: "A New Determination of Molecular Dimensions." (Einstein 1905b; dated Bern, 30 April 1905; dissertation submitted to the University of Zurich.)

11 May: "On the Movement of Small Particles Suspended in Stationary Liquids Required by the Molecular-Kinetic Theory of Heat." (Einstein 1905c; dated Bern, May 1905; received on 11 May 1905 by the *Annalen der Physik.*)

30 June: "On the Electrodynamics of Moving Bodies." (Einstein 1905d; dated Bern, June 1905; received on 30 June 1905 by the *Annalen der Physik.*)

This burst of productivity is especially remarkable in light of the fact that Einstein was occupied forty-eight hours a week as a technical expert, third class, at the Bern patent office, a job that he described to Johannes Stark as "eight hours of strenuous work each day."[1] Note that only eleven days separate the completion of the doctoral dissertation and the completion of the Brownian motion paper. Given that Einstein had published nothing over the twelve-month period prior to this,[2] one might imagine that over the intervening period he had been steadily pursuing several lines of inquiry that happened to come to fruition almost simultaneously. Einstein's correspondence with Michele Besso suggests that he was already at work on the basic strategy of this doctoral dissertation early in 1903.[3]

This appears to have been the pattern of working that Einstein had followed earlier. Following the completion of his first publication, on the consequences of capillarity for the determination of intermolecular forces in liquids, in December of 1900 (Einstein 1901), Einstein's second publication, a similar investigation involving salt solutions (Einstein 1902a), and his third, on the general molecular theory of heat (Einstein 1902b), were not finished until the end of April and end of June 1902, respectively.

By the end of 1901 he had also written and submitted to Professor Alfred Kleiner of the University of Zurich a doctoral dissertation on intermolecular forces in gases, which was later either withdrawn or rejected. Einstein's letters from this period suggest that he was at work on all three of these projects over the course of the year 1901. On 14 April he wrote to Marcel Grossmann, "I shall utilize all the already existing [experimental] results [of others] in my doctoral dissertation" (Einstein 1987, Doc. 100, p. 290). The next day, in a letter to Marić, he discussed his "idea for the investigation of salt solutions," and suggested to her that he restrict himself to the case of infinitely dilute solutions (Einstein 1987, Doc. 101, p. 292). In early September he informed Grossmann:

> Lately I have been engrossed in Boltzmann's works on the kinetic theory of gases and this last few days I wrote a short work myself that provides the keystone in the chain of proofs that he had started. I'll probably publish it in the *Annalen*. (Einstein 1987, Doc. 122, p. 315)

The description of this "short work" prefigures the 1902 paper on the general molecular theory of heat.[4]

In the case of the genesis of special relativity, however, there is *prima facie* reason to think that Einstein arrived at the body of results presented in "On the Electrodynamics of Moving Bodies," in a sudden burst of theoretical creativity, only after he had completed his first three works in the spring of 1905. For the key insight—the discovery of the relativity of simultaneity—occurred to Einstein only late that May after the completion of the Brownian motion paper. The sources documenting this, although indirect and problematic in various regards, are independent and concur in portraying that discovery as a sudden breakthrough that immediately resolved a "problem" that had vexed Einstein to the point of deep frustration. In a biography published under the pseudonym "Anton Reiser," Einstein's son-in-law, Rudolph Kayser, reports:

> Through many long years of hope and disappointment Albert carried on his experiments till he reached the final solution to his problem. When he held the key, with which he was to open the closed door, he despaired and said to his friend[5] "I'm going to give it up."
>
> But next day it was in the greatest excitement that he took up his duties at the office. He could apply himself to the dull routine only with effort. Feverishly he whispered to his friend that now at last he was on the right track. He had made the revolutionary discovery that the traditional conception of the absolute character of simultaneity was a mere prejudice, and that the velocity of light was independent of the motion of coördinate systems. Only five weeks elapsed between this discovery and the first formulation of the special theory of relativity in the treatise entitled "Towards the Electrodynamics of Moving Bodies." (Reiser 1930, pp. 68–69)

According to notes taken by Jun Ishiwara on a lecture that Einstein gave at Kyoto University in 1922, Einstein related the following story:

> By chance a friend of mine in Bern helped me out. It was a beautiful day when I visited him with this problem. I started the conversation with him in the following way: "Recently I have been working on a difficult problem. Today I come here to battle against that problem with you." We discussed every aspect of this problem. Then suddenly I understood where the key to this problem lay. Next day I came back to him again and said to him, without even saying hello, "Thank you. I've completely solved the problem." An analysis of the concept of time was my solution. Time cannot be absolutely defined, and there is an inseparable relation between time and signal velocity. With this new concept, I could resolve all the difficulties completely for the first time.
>
> Within five weeks the special theory of relativity was completed. (Einstein 1922a, p. 46)

Similarly, the gestalt psychologist Max Wertheimer reports that Einstein had indicated to him in 1916 that "from the moment, however, that he came to question the customary concept of time, it took him only five weeks to write his paper on relativity—although at this time he was doing a full day's work at the Patent Office" (Wertheimer 1945, p. 214).

Einstein later revealed that the problem that had stumped him at the time arose from a paradox that had already occurred to him at the age of sixteen:

> If I pursue a beam of light with the velocity *c* (velocity of light in a vacuum), I should observe such a beam of light as an electromagnetic field at rest though spatially oscillating. There seems to be no such thing, however, neither on the basis of experience nor according to Maxwell's equations. From the very beginning it appeared to me intuitively obvious that, judged from the standpoint of such an observer, everything would have to happen according to the same laws as for an observer who, relative to the earth, was at rest. For how should the first observer know, or be able to determine, that he is in a state of fast uniform motion? (Einstein 1979, pp. 50–51)

Einstein amplified:

> The above paradox may then be formulated as follows. According to the rules of connection, used in classical physics, between the spatial coordinates and the time of events in the transition from one inertial system to another, the two assumptions of
>
> (1) the constancy of the light velocity
> (2) the independence of the laws (thus especially also of the law of the constancy of the light velocity) from the choice of inertial system (principle of special relativity)
>
> are mutually incompatible (despite the fact that both taken separately are based on experience). (Einstein 1979, p. 53)

In other words, calling the first of these assumptions the *Light Postulate* and the second simply the *Principle of Relativity*, the Light Postulate and the Principle of Relativity are jointly incompatible with the classical velocity addition rule. Time and time again in expounding the theory of special relativity, Einstein had motivated the need for the Lorentz transformations in just this fashion. It would seem, then, that in order to understand the genesis of the theory it suffices to explain how Einstein came to have enough confidence in the Light Postulate and the Principle of Relativity to call into question the validity of the Galilean transformations. What do we know about Einstein's attitude toward these two principles over the period of time leading up to May of 1905?

3. Aether and Motion

First, a point of clarification concerning the Principle of Relativity. This principle should be distinguished from a weaker one that I shall call the *null result thesis*, which maintains only that there are no observable phenomena that can distinguish a system in absolute motion from one at rest. The Principle of Relativity requires that the concept of absolute rest should play no role in the formulation of the basic laws of physics, while the null result thesis makes no necessary prohibition to this effect. The former principle entails the latter but not conversely, although one might construe the null result thesis as inductive warrant for the Principle of Relativity, as did Einstein in the 1905 relativity paper.

There is a further subtlety to the interpretation of the Principle of Relativity in regard to the content of the concept of absolute rest. H.A. Lorentz was prepared to fight off the charge that his version of electrodynamics as presented in the *Versuch* committed him to accepting the validity of that concept:

> That there can be no question concerning the *absolute* rest of the aether is probably self-evident; the expression would not even have a meaning. If for brevity's sake I say that the aether is at rest, it is thereby meant only that one part of this medium is not displaced with respect to another and that all observable motions of the heavenly bodies are relative motions with respect to the aether. (Lorentz 1895, p. 4)[6]

On the other hand, the primary function of Lorentz's de-materialized aether can be seen to be that of singling out a preferred frame of reference and thus of providing a criterion of absolute motion in terms of relative motion with respect to that frame (much as Newton defined absolute motion as motion relative to absolute space).

The issue is even murkier for those versions of optics and electrodynamics that assume that the aether partakes in the motion of material bodies. For then the aether formally resembles a fluid medium such as air or water, and it is rather unnatural to hold that the laws of acoustics or hydrodynamics violate the Principle of Relativity.[7] Nonetheless, since the great bulk of the aether in interstellar space remains relatively quiescent, it can be maintained, as it was by August Föppl in his textbook on Maxwell's theory, that "the absolute motion of [a] body is to be understood as that which the body executes relative to the uninvolved and distant aether" (Föppl 1894, p. 309).

In dismissing the aether in the 1905 relativity paper, Einstein alluded to both its stationary and dragged versions:

> The introduction of a "light aether" will prove superfluous inasmuch as in accordance with the concept to be developed here, no "space at absolute rest" endowed with special properties will be introduced, nor will a velocity vector be assigned to a point of empty space at which electromagnetic processes take place. (Einstein 1905d, p. 892; 1989, p. 277)

A good number of Einstein's later expositions begin with a discussion of the drawbacks of a movable aether and cite Fizeau's experiment on the velocity of light in moving water as a crucial experiment.[8] It appears that the choice between a movable and a completely stationary aether dominated his earliest thinking on the electrodynamics of moving bodies. Föppl, whom the biographies of both Reiser and Philipp Frank list as one of the authors whom Einstein had read as a student,[9] singles out the resolution of the question whether or not the aether is dragged as "perhaps the most important task of contemporary research" (Föppl 1894, § 114). That the problem addressed by Fizeau's experiment was an issue for Einstein in his student days is borne out by a letter to Marić written in early September 1899:

> A good idea occurred to me in Aarau for investigating what influence the relative motion of bodies with respect to the light aether has on the velocity of propagation of light in transparent bodies. Also a theory on this matter has come to my mind, which seems to me to have a high probability. (Einstein 1987, Doc. 54, p. 230)

Later that month he informed Marić that he had written to Wilhelm Wien about the problem:

> I also wrote to Professor Wien in Aachen about the paper on the relative motion of the luminiferous aether against ponderable matter, which the "Principal"[10] treated in such a stepmotherly fashion. I read a very interesting paper published by this man on the same topic in 1898. (Einstein 1987, Doc. 57, pp. 233–234)

That article is titled "On Questions that Relate to the Translatory Motion of the Light Aether" (Wien 1898). In it Wien reviews the difficulties facing both the movable and stationary conceptions of the aether, listing, with brief descriptions, thirteen experimental results, including those of Fizeau and Michelson-Morley. Interestingly enough, Einstein appears to have been already disposed against a movable aether. The preceding August he had written to Marić:

> I am more and more convinced that the electrodynamics of moving bodies, as presented today, is not correct, and that it should be possible to present it in a simpler way. The introduction of the term "aether" into the theories of electricity led to the notion of a medium of whose motion one can speak without being able, I believe, to associate a physical meaning with this statement. (Einstein 1987, Doc. 52, p. 226)

This passage is certainly suggestive. If Einstein had found a movable aether objectionable on the grounds that it might not be possible to assign a "physical meaning" to the motion of that medium, would he have found a stationary aether any more acceptable unless there were some way to detect the motion of material bodies with respect to it? Both Reiser's biography and Ishiwara's notes of the Kyoto lecture mention proposals that Einstein made in his student days to investigate experimentally the motion of the earth through the aether. Some of his correspondence from a few years later suggests the same. In September of 1901 he wrote to Grossmann: "A considerably simpler method of investigating the relative motion of matter with respect to the luminiferous aether that is based on ordinary interference experiments has just occurred to me" (Einstein 1987, Doc. 122, p. 316). Also the following December he informed Marić

> [Kleiner] advised me to publish my ideas about the electromagnetic theory of light for moving bodies together with the experimental method. He thought that the experimental method proposed by me is the simplest and most appropriate one conceivable. (Einstein 1987, Doc. 130, p. 328)

Although Einstein was aware of Lorentz's theory by this time, if only from the very brief discussion of it in the article by Wien that he had read a few years earlier, he was most likely unfamiliar with its details. Later in December he again wrote to Marić: "I now want to buckle down and study what Lorentz and Drude have written on the electrodynamics of moving bodies.[11] Ehrat must get the literature for me" (Einstein 1987, Doc. 131, p. 330). When he finally did "buckle down" with Lorentz, he was truly impressed, as his later repeated expressions of admiration for Lorentz's electrodynamics attest. Despite this and Lorentz's disclaimers about absolute motion, Einstein eventually came to be haunted by the fact that

> the theory gives rise to an, as it were, spooky feature: namely, while the equations of mechanics appear to have a claim to validity in each inertial

> frame, the validity of Maxwell's equations is restricted to a coordinate system in a completely determined state of motion; it appears to presuppose an absolutely stationary physical space (stationary aether), in spite of the fact that the mechanical aether theory had just been abandoned. This was all the more remarkable, since all attempts to demonstrate the motion of the earth with respect to this "aether" by means of optical and electromagnetic experiments had turned out negative.[12]

Thus, despite the virtues he found in Lorentz's theory, he could not rest entirely content with it, having in the meanwhile come to adopt the Principle of Relativity. Before considering when this might have occurred, let me turn to Einstein's attitudes toward the light postulate in the years leading up to 1905.

4. The Light Postulate

The Light Postulate is easily confused with the stronger statement that the velocity of light is a fixed constant *c* in all inertial frames. As a consequence, numerous textbooks and commentaries cite the various "null result" experiments, in particular the Michelson-Morley experiment, as evidence for the Light Postulate. In the 1905 relativity paper, however, "the failure of attempts to detect a motion of the earth relative to the 'light medium'" is used as evidential support only for the Principle of Relativity. (The Light Postulate is introduced, almost parenthetically, without any discussion of experimental grounds.) As John Stachel has emphasized, when Einstein cited the Michelson-Morley experiment in subsequent expositions of special relativity, he did so exclusively for this purpose (Stachel 1982). The reason is that, although the familiar null results can be marshaled to call into doubt the existence of a stationary aether, the Light Postulate is a necessary consequence of the wave theory of light, and hence of Lorentz's theory, since the velocity of propagation with respect to the aether is a fixed constant independently of whether or not the light source is in motion. In fact, Einstein often cited the experimental success of Lorentz's theory in support of the Light Postulate:

> The development of theoretical physics shows, however, that we cannot pursue this course [viz., abandonment of the Light Postulate]. The epoch-making theoretical investigations of H.A. Lorentz on the electrodynamical and optical phenomena connected with moving bodies show that experience in this domain leads conclusively to a theory of electromagnetic phenomena, of which the law of the constancy of the velocity

> of light *in vacuo* is a necessary consequence. Prominent theoretical physicists were therefore more inclined to reject the principle of relativity in spite of the fact that no empirical data had been found which were contradictory to this principle. (Einstein 1917, p. 19)

Before 1905, however, Einstein found himself inclined to follow the path investigated by a number of physicists only after 1905 in an attempt to avoid the new doctrine of space and time contained in the special theory of relativity,[13] namely, to reject the Light Postulate in favor of what subsequently came to be known as the emission hypothesis—that the velocity of light is *c* only in the inertial frame occupied by the light source at the moment of emission. As Robert Shankland reports from his interviews with Einstein in the early fifties: "This led him to a discussion of emission theories of light, and he told me he had thought of and abandoned the (Ritz) emission theory before 1905" (Shankland 1963, p. 49). This is confirmed by a number of letters. For example, Einstein wrote to Mario Viscardini in April of 1922:

> The hypothesis articulated in the article, that in free space light has the constant velocity *c*, not with respect to the coordinate system, but relative to the light source, was discussed for the first time in detail by the Swiss physicist W. Ritz and was seriously taken into consideration by myself before the formulation of the special theory of relativity. (EA 25-302)[14]

To his frustration, though, Einstein found himself unable to construct a plausible emission theory. As he later wrote in a draft of a letter to A.P. Rippenbein in 1952:

> Moreover this theory requires that everywhere and in each fixed direction light waves of a different velocity of propagation should be possible. It may well be impossible to set up an electromagnetic theory that is in any way reasonable and accomplishes such a feat. This is the principal reason why, even before the formulation of the special theory of relativity, I rejected this way out, which is intrinsically conceivable. (EA 20-046)[15]

Thus, having already convinced himself of the Principle of Relativity, Einstein found himself without any reasonable alternative to the Light Postulate, even though it appeared to stand in direct contradiction to the Principle of Relativity. If we imagine that he arrived at this dilemma in May of 1905, then it would appear that we have a reasonably satisfactory picture of the emergence of the theory: The recognition of the relative character of distant simultaneity indicated that neither of these principles was defective, but rather that the defect lay in the Galilean transformations, which he had previously taken for granted. Given this, he was able to establish an alternative set of transformation equations, namely, the Lorentz transformations, which rendered the two principles compatible. On this basis Einstein worked out the new kinematics as presented in Part I of the 1905 paper. One might even surmise that he discovered the Lorentz transformations by deriving them more or less according to the method given there. Only after having this kinematics in hand, the story continues, did Einstein investigate the consequences for optics and electrodynamics, eventually obtaining the results that appear in Part II of the paper.

5. The Electromagnetic Asymmetries

Although short on detail, this tidy picture might seem to be correct at least in outline. It might, however, be just *too* tidy.

As Gerald Holton remarked, "Einstein's published papers in their architectural details do not necessarily correspond, point for point, with the sequence of his actual thought processes in arriving at his conclusions" (Holton 1980, p. 64). Anyone who has worked through the derivation of the Lorentz transformations presented in the 1905 paper will suspect as much. There is reason, however, to think that what Holton says is true not only of the finer details of the blueprint, but of the larger outline as well. The gestalt psychologist, Max Wertheimer, conducted interviews with Einstein beginning in 1916 with the specific purpose of finding material for a case study in the psychology of scientific discovery. Although Wertheimer's reconstructed account of the discovery of special relativity should be read with caution, Einstein's voice can be clearly heard at various points, in particular in the following lengthy footnote:

> In this respect I wish to report some characteristic remarks of Einstein himself. Before the discovery that the crucial point, the solution, lay in the concept of time, more particularly that of simultaneity, axioms played no role in the thought process—of this Einstein is sure. (The very moment he saw the gap, and realized the relevance of simultaneity, he knew this to be the crucial point for the solution.) But even afterward, in the final five weeks, it was not the axioms that came first. "No really productive man thinks in such a paper fashion," said Einstein. "The way the two triple sets of axioms are constructed in the Einstein-Infeld book is not at all the way things happened in the process of actual thinking. This was merely a later formulation of the subject matter, just a question of how the thing could afterwards best be written. The axioms express essentials in a condensed form. Once one has found such things, one

enjoys formulating them in that way; but in this process they did not grow out of any manipulation of axioms." (Wertheimer 1945, p. 229)

A manuscript from around 1920, brought to public attention by Holton, provides a clue to the sort of further factors that might be taken into consideration.

> In the development of special relativity theory, a thought—not previously mentioned—played for me a leading role.
>
> According to Faraday, when a magnet is in relative motion with respect to a conducting circuit, an electric current is induced in the latter. It is all the same whether the magnet moves or the conductor: only the relative motion counts, according to the Maxwell-Lorentz theory. However, the theoretical interpretation of the phenomenon in these two cases is quite different: If it is the magnet that moves, there exists in space a magnetic field that changes with time and which, according to Maxwell, generates a closed line of electric force—that is, a physically real electric field: the electric field sets in motion movable electric masses within the conductor.
>
> However, if the magnet is at rest and the conducting circuit moves, no electric field is generated: the current arises in the conductor because the electric bodies being carried along with the conductor experience an electromotive force, as established hypothetically by Lorentz, on account of their (mechanically enforced) motion relative to the magnetic field.
>
> The thought that one is dealing here with two fundamentally different cases was, for me, unbearable. The difference between these two cases could not be a real difference but rather, in my conviction only a difference in the choice of reference point. Judged from the magnet, there certainly were no electric fields; judged from the conducting circuit, there certainly was one. The existence of an electric field was therefore a relative one, depending on the state of motion of the coordinate system being used, and a kind of objective reality could be granted only to this *electric and magnetic field together*, quite apart from the state of relative motion of the observer or the coordinate system. The phenomena of electromagnetic induction forced me to postulate the (special) principle of relativity. Footnote: The difficulty that had to be overcome was in the constancy of the velocity of light in a vacuum, which I had first thought I would have to give up. Only after groping for years did I notice that the difficulty rests on the arbitrariness of the fundamental kinematical concepts. (EA 2-070)[16]

The 1905 paper begins with this very example of the magnet and conductor, but only as a familiar example in which the phenomena depend only on relative motion. The novel insight that the electric and magnetic fields have only a relative existence is not mentioned and developed until the second part of the paper after the presentation of the new kinematics, even though it offers, by itself, a potentially powerful heuristic for the investigation of specific problems in optics and electrodynamics, comparable to that of the Principle of Equivalence for gravitational theory. The centrality of its role is borne out by Einstein's later comment that: "My *direct* [emphasis mine] path to the special theory of relativity was determined above all by the conviction that the induced electromotive force in a conductor moving in a magnetic field was nothing else but an electric field" (draft of a letter from Einstein to Shankland, 1952; as quoted in Holton 1969, p. 339, n. 70; translation mine). This suggests that the specific physical applications treated in the 1905 paper find their way into that paper because of their intrinsic interest to Einstein in the genesis of the theory and not simply because of their usefulness in illustrating consequences of the theory.

6. The Thematic and Logical Structure of the 1905 Paper

To get a fix on these problems, let me briefly review the structure and contents of the 1905 relativity paper (see Figure 1, p. 14). Figure 1, apart from emphasizing the contents of the three introductory paragraphs, is simply a section-by-section table of contents of the paper. However, I have boxed those sections that belong together in terms of their subject matter. The grouping together of the kinematical sections deserves no comment. Sections 7 and 8 belong together, at least superficially, on the grounds that it is in these sections alone that Einstein is concerned with relativistic properties of radiation beyond simply the invariance of c. Shortly, we will see a deeper rationale for this grouping.

What I want to draw particular attention to is the grouping together of the initial paragraph with the first and last sections (§6 and §10) of the "Electrodynamical Part." The opening paragraph, as already mentioned, develops the case of the magnet and moving conductor quoted above. Section 6 derives the field transformation equations from the covariance of the Maxwell-Hertz equation for free space and takes up again the general problem illustrated in the opening paragraph, namely, the motion of a point charge in a magnetic field. According to the principle of relativity, the field quantities are frame dependent. The force acting on the charge according to the "new conception" is due solely to the electric field in the rest frame of the charge at its location, and thus the electromotive force(proportional "to first order" to $\mathbf{v} \times \mathbf{H}$), which according to the "customary conception" acts on a moving charge, has merely the status of an "auxiliary concept."

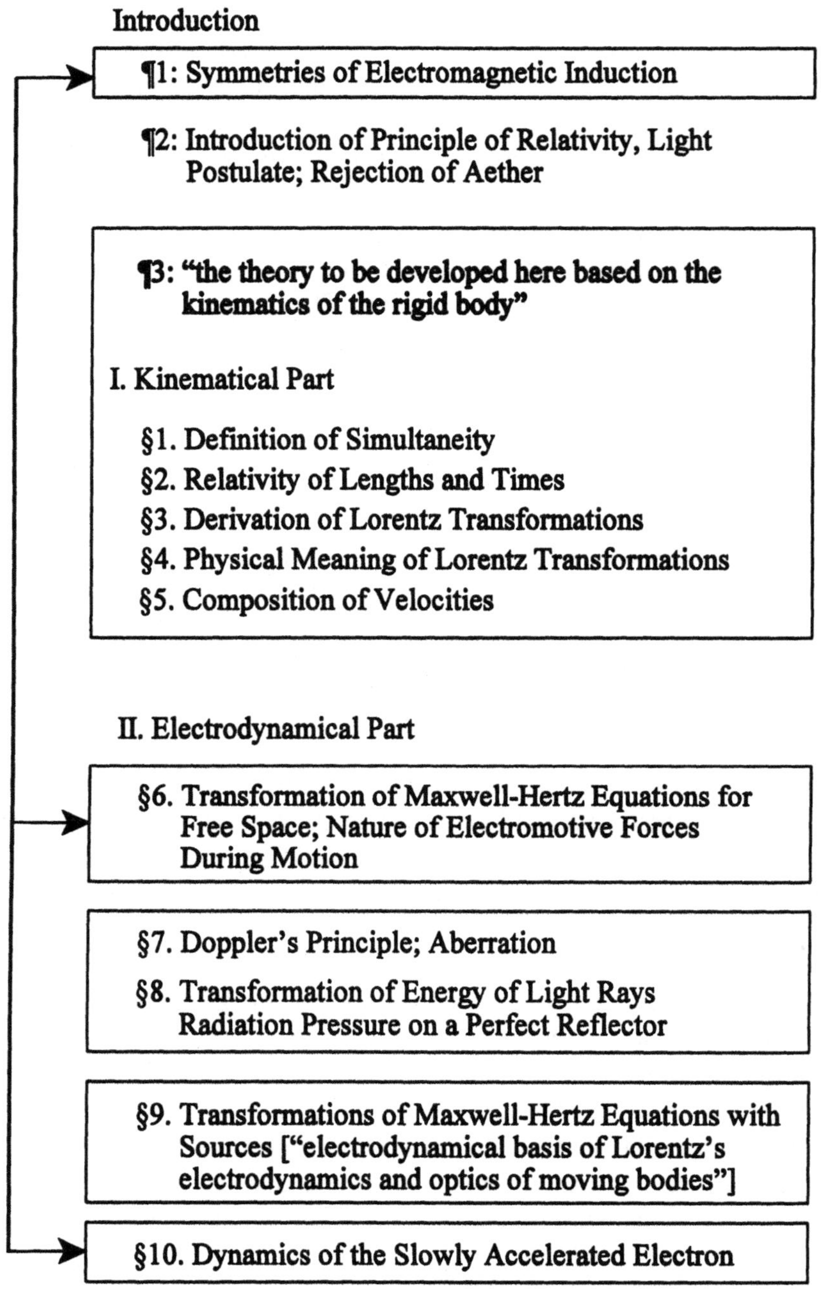

Figure 1

Thematic Structure of "On the Electrodynamics of Moving Bodies"

Section 10, with which the paper closes, then completes the dynamical treatment of a moving charge.[17] Einstein considers here a point charge with (rest) mass μ moving in an electromagnetic field and uses the "new conception," according to which the force acting on the electron is given by the electric field relative to its rest frame, in order to derive expressions for the "longitudinal" and "transverse" mass of the electron, the kinetic energy of the electron, the ratio of electric to magnetic deflection, the relation between velocity and electrostatic potential difference traversed, and the radius of curvature of the path in a magnetic field. Of the physical matters addressed in the paper, the problem of the motion of a point charge in an electromagnetic field without question plays a distinguished role.

It might be wondered why section 9, which treats the field equations with convection currents, has been separated from the rest. After all, a convection current is just a current of moving charges. However, the influence of an electromagnetic field on convection currents is not addressed in this section. Moreover, the conception of the "electron" here is not the point electron of section 10, but that of a small rigid body to which a charge density is coupled. The purpose of section 9 is to establish that the field equations with convection terms included conform to the Principle of Relativity. The upshot, as Einstein makes clear, is to demonstrate "that on the basis of our kinematical principle the electromagnetic foundation of Lorentz's theory of the electrodynamics of moving bodies is in accord with the Principle of Relativity" (Einstein 1905d, p. 917). Although it goes unsaid, it follows immediately from this that the great successes of Lorentz's theory in explaining the optical properties of material media, including the Zeeman effect and the propagation of light in moving media, are not undermined by the Principle of Relativity.

A detailed look at the logical structure of the paper will, I think, further illuminate the nexus of Einstein's concerns. In Figure 2 (on p. 17), I have placed the Principle of Relativity, the Light Postulate, and the definition of simultaneity at the fountainhead of the chart. Immediately branching off from this, without requiring the Lorentz transformations, are qualitative kinematical consequences, namely, the relativity of simultaneity and the need to distinguish between the geometric and kinematical conceptions of the shape of a rigid body. The quantitative treatment of the latter, together with the "time dilation" of a moving clock, depends on the derivation of the Lorentz transformations from the fundamental postulates and the definition of simultaneity. The relativistic velocity addition rule plays two roles—to show the group character of the Lorentz transformations and to secure the compatibility of Lorentz's electron theory with the Principle of Relativity.

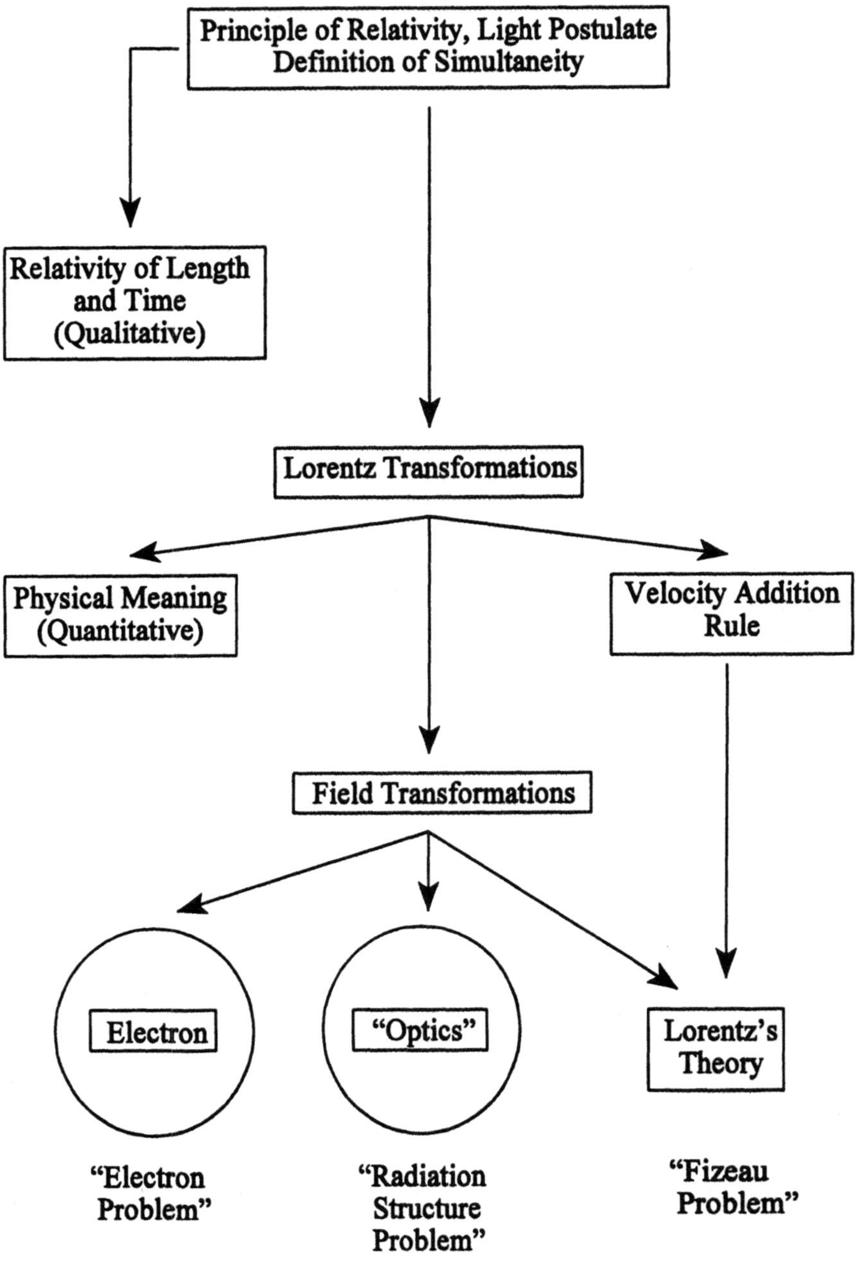

Figure 2

Logical Structure of "On the Electrodynamics of Moving Bodies"

The field transformation equations are obtained by requiring the covariance of the Maxwell-Hertz equations for free space under the Lorentz transformations. It should be noted that the Lorentz transformations follow in a similar fashion by requiring the covariance of these equations under the field transformation equations. Joint application of the field transformation equations and the Lorentz transformations then suffice for a complete (neglecting radiation) mathematical solution of the dynamics of a charged point mass—what I will call "the electron problem."

Joint application of the Lorentz and field transformations also suffice for treatment of the properties of radiation in sections 7 and 8. Here, a fine-grained look at the interrelations of the results gives some clue as to the nature of Einstein's concerns. Einstein begins by applying the Lorentz and field transformations to the equation for a linearly polarized plane wave. From this follow equations for the Doppler shift in frequency, the angle of aberration, and the amplitude of the wave in a moving frame. The latter, taking into account the frame dependence of the volume of a portion of the wave front, permits us to calculate the ratio of the total radiation energy in a moving frame to that in the "stationary" frame. The key result here is that this is the same as the ratio of the frequencies given by the Doppler effect. Thus the ratio of total energy to frequency of radiation is frame independent. Finally, Einstein applies the equations for the Doppler shift in frequency, the angle of aberration, and the transformation of wave amplitude together to calculate the pressure of radiation on a moving mirror.

There is a tendency on our part to associate these applications to radiation with the traditional attempts to explain away the "null results" of the efforts to measure the velocity of the earth with respect to the aether. Such topics as aberration, the Doppler shift, and even radiation pressure are treated in Lorentz's *Versuch*, and the *Versuch* is now most often remembered in connection with Lorentz's result that the motion of the earth can have no first-order influence on optical phenomena. However, nothing more than the Principle of Relativity is required to explain these null results, and, although Doppler's principle and aberration did occupy niches in contemporary discussions of the optics and electrodynamics of moving bodies, the logical structure of sections 7 and 8 suggests that Einstein had reasons other than mere tradition for focusing on the results he did. The formulas for aberration and Doppler's principle from section 7 serve as lemmas for the two outstanding results of section 8:

(i) the invariance of the ratio of energy per unit volume to frequency, and

(ii) the calculation of the pressure of radiation exerted on a moving mirror.

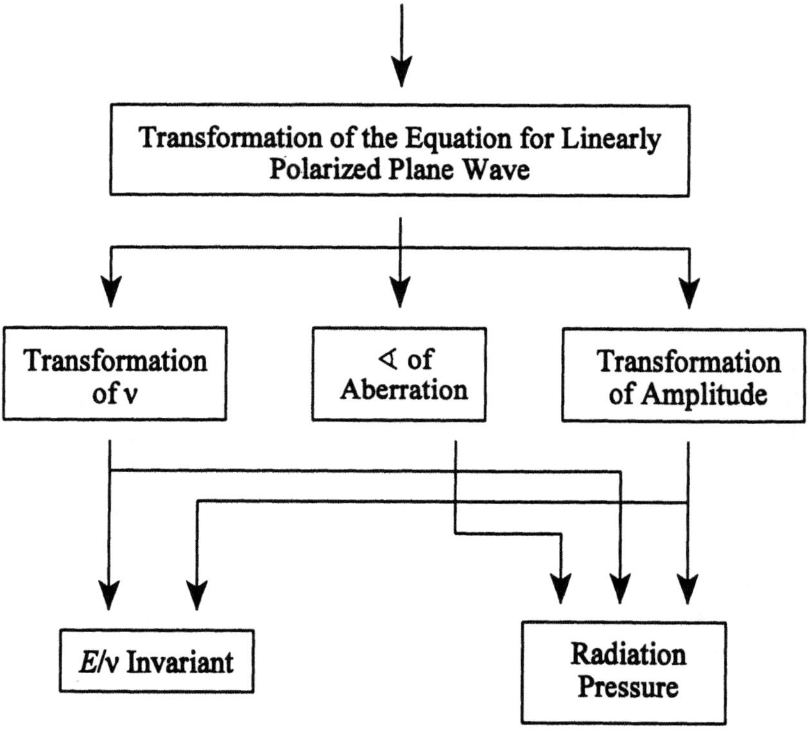

Figure 3

"Radiation–Structure" Problem

Both of these were of obvious interest to Einstein in connection with his concerns about the microstructure of radiation. The first of the above results indicates that, in the "Wien regime," that is, in the high frequency, low density region of thermal radiation for which the Wien distribution law holds, the number of "light quanta" per unit volume is a relativistic invariant. Without this result, it is impossible to attribute an objective reality to such light quanta. The second result is also associated with the "blackbody" problem. Radiation pressure and the characteristics of reflected radiation played an integral role in the history of theoretical investigations of the thermodynamics of blackbody radiation and its spectrum—in Ludwig Boltzmann's derivation of the Stefan-Boltzmann law (Boltzmann, 1884), in Wien's derivation of his displacement law (Wien 1893, 1894), and in later studies by Max Abraham and Fritz Hasenöhrl (Abraham 1904a, 1904b; Hasenöhrl 1904). In addition, there is Einstein's own professed concern as to the inability of Maxwell's equations to account for the pressure fluctuations on a moving mirror. Einstein wrote to Max von Laue on 17 January 1952:

> When one goes through your collection of verifications of the Special Relativity Theory, one believes that Maxwell's Theory is firmly established. But in 1905 I knew already with certainty that it leads to the wrong fluctuations in radiation pressure, and consequently to an incorrect Brownian motion of a mirror in a Planckian radiation cavity. (EA 16-167).[18]

In his *Autobiographical Notes*, Einstein identified two "fundamental crises" in the development of physics as he saw it in the first years of this century. The first of these was the task of completing the so-called "electromagnetic world view," a project to replace mechanics with electrodynamics as the foundation for all of physics. The "second fundamental crisis" involved the difficulty of finding an explanation of the distribution law for blackbody radiation. The electron problem was an integral feature of the first of these crises, and Einstein's concerns with the microstructure of radiation were intimately associated with the second. Undoubtedly, his interests in these arose before the spring of 1905 and undoubtedly he thought hard about them.

7. The Einstein Circle

Einstein's response to the electromagnetic world view and the electron problem was determined by his reaction to the symmetries of Faraday induction and the conception of the relativity of the electric and the magnetic field, resulting in the adoption of the Principle of Relativity and the rejection of the electromagnetic aether. His eventual response to the radiation-structure problem was to propose that the energy of radiation in free space in the Wien regime "consists of a finite number of energy quanta that are localized at points in space, move without dividing, and can be absorbed or generated only as a whole" (Einstein 1905a, p. 133).

Although Einstein discussed the background to these two "fundamental crises" separately in the *Autobiographical Notes*, his responses to them are intimately linked in a circle of ideas (see Figure 4). We have seen how Einstein initially inferred that the Principle of Relativity demands that the velocity of light depend on the motion of the source. This assumption naturally brings to mind the corpuscular doctrine of light historically associated with it. There is another line of reasoning that also suggests that

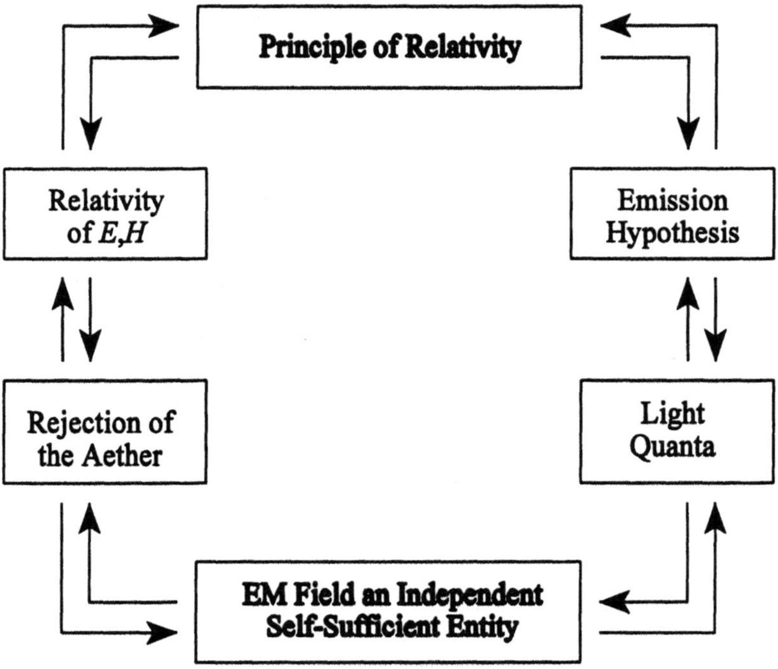

Figure 4

Circle of Ideas

radiation might ultimately have a particulate structure. Given the Principle of Relativity and the realization of the relative character of the field quantities, the aether becomes entirely superfluous. As Einstein explained in an address to the Deutsche Physikalische Gesellschaft in 1909: "In that case the electromagnetic fields that constitute light will no longer appear to be states of a hypothetical medium, but rather independent entities emitted by the sources of light, exactly as in the Newtonian emission theory of light" (Einstein 1909b, p. 487). Conversely, the idea of light quanta leads us back to the Principle of Relativity either through the rejection of the aether or through the consideration of the emission hypothesis.

There are many ways in which Einstein might conceivably have broken into this circle of ideas. Besides his failure to find anywhere a demonstration of the motion of the earth and his reflections on the symmetries of Faraday induction, a host of considerations might have suggested—either individually or together—the hypothesis of light quanta.

First, there is the experimental data on photoluminescence, the photoelectric effect, and the ionization of gases. Second, Einstein may simply have wondered to what extent his "general molecular theory of heat," developed for systems with a finite number of degrees of freedom, might be applicable to radiation as well.[19] Third, it may have occurred to Einstein that all elementary processes, including those involving radiation, should be completely reversible (this is perhaps a specific aspect of the last consideration). As he explained in 1909:

> While in the kinetic theory of matter there exists an inverse process for every process in which only a few elementary particles take part, e.g., for every molecular collision, according to the wave theory this is not the case for elementary radiation processes. According to the prevailing theory, an oscillating ion produces an outwardly propagated spherical wave. The opposite process does not exist as an *elementary process*. It is true that the inwardly propagated spherical wave is mathematically possible; however, its approximate realization requires an enormous number of emitting elementary structures. Thus, the elementary process of light radiation as such does not possess the character of reversibility. Here, I believe, our wave theory is off the mark. Concerning this point the Newtonian emission theory of light seems to contain more truth than does the wave theory, since according to the former the energy imparted at emission to a particle of light is not scattered throughout the infinite space but remains available for an elementary process of absorption.(Einstein 1909b, p. 491)[20]

Finally, Einstein may have inferred that radiation has a particulate structure from a derivation of the so-called "ultraviolet catastrophe," that is, the result that the equipartition theorem applied to a system of oscillators in thermal equilibrium with a gas, together with Planck's expression for the mean energy of an oscillator as derived from Maxwell's theory, leads to a blackbody radiation distribution function whose integral over all frequencies diverges.

Conversely, these are insights or conclusions that Einstein might equally well have carried away from the circle of ideas. It should also be noted that certain links in the circle appear stronger than others, which may make a difference in judging the relation of the circle to the actual historical events. It strikes me that much more is required to get from the Principle of Relativity and the nexus of ideas more intimately associated with it to the idea of light quanta, than from the idea of light quanta to the rejection of the aether and the adoption of the Principle of Relativity. A number of physicists after 1905, beginning with Ritz, attempted to construct emission theories conforming to the Principle of Relativity without thereby being led to the light quantum hypothesis. There is, however, strong cause to suppose that by the time Einstein began to seriously consider light quanta, he either was already committed to the Principle of Relativity or else committed himself to the principle shortly thereafter. When, then, did the possibility of light quanta occur to him?

8. The Radiation-Structure Problem

Einstein's second paper on statistical thermodynamics (Einstein 1903) sought to derive "the foundations of the theory of heat" from a few general assumptions. One of these was that the state of a physical system is uniquely determined by a finite number of scalar quantities. His next publication (Einstein 1904), completed in March of 1904, presents, as he put it, "a few addenda" to the previous paper. Noteworthy among these is the derivation of the energy fluctuation formula

$$\langle \varepsilon^2 \rangle = 2\kappa T^2 \frac{d\langle E \rangle}{dT},$$

which relates the measure of the thermal stability of the system

$$\langle \varepsilon^2 \rangle =_{df} \langle E^2 \rangle - \langle E \rangle^2$$

to the "absolute" constant κ (one half the value of Boltzmann's constant). Einstein proceeds to apply this formula together with the Stefan-Boltzmann law to a system of blackbody radiation, deriving "without recourse to special hypotheses," the Wien displacement law

$$\lambda_{max} T = a$$

with a value for the constant a in rough agreement with experiment. Since one of the premises in the derivation of the energy fluctuation formula is that the system in question has a finite number of degrees of freedom, we might ask:

Query: Is the derivation of the Wien displacement law in Einstein 1904 premised on the assumption that a "radiation space" is a system with a finite number of degrees of freedom?

In answering this, one must weigh various remarks in that paper against one another. The comment that the energy fluctuation formula "is interesting because it no longer contains any quantity reminiscent of the assumptions on which the theory is based" suggests that Einstein's purpose in applying the fluctuation formula to blackbody radiation is to investigate whether the fluctuation formula might apply even to systems that do not satisfy the assumptions of his so-called molecular theory of heat. Alternatively, this remark might be taken to signal only that the formula can be applied without having to know anything about the dynamical details of the system in question, assuming that it is one of the kind meeting the assumptions of the molecular theory of heat. Einstein at least called the reader's attention to the possibility that a "radiation space" is a system of this kind, and hence one with a finite number of degrees of freedom:

> Of course, one can object that we are not permitted to assert that a radiation space should be viewed as a system of the kind we have assumed, not *even if the applicability of the general molecular theory is conceded.* Perhaps one would have to assume, for example, that the boundaries of the space vary with its electromagnetic states. (Einstein 1904, p. 361; emphasis mine)

In the conclusion of his discussion of blackbody radiation in this paper, Einstein suggests that the agreement between the derived formula and experiment should be construed as evidence for the applicability of the general molecular theory in this instance:[21]

> One can see that not only the manner of the dependence on the temperature but also the order of magnitude of λ_m can be correctly determined *by means of the general molecular theory of heat*, and I believe that given the broad generality of our assumptions, this agreement should not be ascribed to chance. (Einstein 1904, p. 362; emphasis mine)

An affirmative answer to the above query would indicate that Einstein had entertained the notion of light quanta, in some form or other, by early 1904. Assuming that this involved an abandonment of the aether and a correlative recognition of the relativity of the field quantities, it provides an upper bound for the dating of Einstein's commitment to the Principle of Relativity. Since the latter initially signaled to Einstein that the velocity of light should depend on the motion of the source, it is natural to imagine that he considered this possibility at the time. In that case, the significance of the derivation of the displacement law from the general molecular theory of heat would appear to be this. Wien's theoretical derivation of the law involved purely thermodynamic considerations in conjunction with specific consequences for the characteristics of radiation reflected from a moving reflector as derived from Maxwell's theory (Wien 1893, 1894). On the emission hypothesis, though, it is not at all evident what these characteristics should be. For example, it is unclear whether the velocity of the reflected radiation in the rest frame of the source should be $c - v$, c, or $c + v$, where v is the relative velocity of the mirror. These possibilities arise by considering the effective source of the reflected radiation to be, respectively, the surface of the mirror, the original source, or the mirror image of the original source. Thus, the question arises whether Maxwell's theory is necessary for a theoretical understanding of the displacement law. The derivation in Einstein's 1904 paper shows that it is not. It shows furthermore that the result follows for a discontinuous radiation field without the need for any specific assumptions regarding the relation of the velocity of propagation to the motion of the source.

Einstein later reported a number of difficulties that he encountered in his attempts to forge an emission theory, and many of these are of the sort that might easily spill over into his thinking about light quanta. In April of 1922 he wrote to Viscardini: "I rejected this [emission] hypothesis at that time, because it leads to tremendous theoretical difficulties (e.g., the explanation of shadow formation by a screen that moves relative to the light source)" (EA 25-301).[22] In a letter that has been dated June of 1912, he had explained to Ehrenfest how this involves assumptions concerning the dependence of the velocity of radiation emitted by a resonator on the cause of its excitation:

> One can adduce for the hypothesis of the independence of the velocity of light from the state of motion of the light source perhaps only its simplicity and easy feasibility. As soon as one gives up this hypothesis, then for the explanation of shadow formation, one must no doubt introduce the ugly assumption that light emitted by a resonator depends on the type of excitation (excitation by "moving" radiation or excitation of another type). (Einstein 1993, Doc. 409)[23]

To various correspondents, Einstein cited three other difficulties with the emission hypothesis. First, it would be possible for light emitted from an accelerating source to "back up" on itself.[24] Second, if the velocity of light from a moving source is modified upon passage through a stationary thin film, this would give rise to unexpected interference phenomena.[25] Third, it seemed to Einstein that there should arise the question why the velocity of light emitted from a stationary body should be completely independent of the color.[26] As Einstein explained to Ehrenfest in June of 1912:

> I considered what would be more probable, the principle of the constancy of *c* as demanded by Maxwell's equations, or the constancy of *c* exclusively for an observer who remains with the light source. I opted for the first, because I came to the conviction that all light should be defined by frequency and intensity alone, completely independently of whether it comes from a moving or from a stationary light source. (Einstein 1993, Doc. 409, p. 485)

This leads us to ask when it was that Einstein came to this conviction.

Query: When did Einstein finally abandon consideration of the emission hypothesis?

Internal evidence from his light quantum paper of 1905 suggests that, at the latest, Einstein had given up on the feasibility of any sort of emission theory by March of that year. For in the introduction to that paper, he expresses his conviction in the "time-average" correctness of the usual wave theory of light, and hence in the validity of the light postulate:

> The wave theory of light, which operates with continuous spatial functions, has proved itself splendidly in describing purely optical phenomena *and will probably never be replaced by another theory*. One should keep in mind, however, that optical observations apply to time averages and not to momentary values, and it is conceivable that despite the complete confirmation of the theories of diffraction, reflection, refraction, dispersion, etc., by experiment, the theory of light, which operates with continuous spatial functions, may lead to contradictions with experience when it is applied to the phenomena of production and transformation of light. (Einstein 1905a, pp. 132–133; emphasis mine)

We may be able to surmise an even earlier date should it turn out that the line of reasoning in the light quantum paper implicitly depends on the Light Postulate.

Query: To what extent does the Light Postulate play a constructive role in the 1905 light quantum paper?

There are essentially four lines of inquiry introduced in the 1905 light quantum paper. The first aims to establish the inadequacy of the classical picture (that is, Maxwell's theory together with the kinetic theory) to account, even qualitatively, for the thermodynamics of blackbody radiation. According to the classical picture, there is no state of equilibrium between radiation and matter. The second aims to show that, despite this, the classical picture gives adequate quantitative results in the limit of long wavelengths and high energy densities, and thus can be used in conjunction with Planck's phenomenologically determined formula to fix the value of Avogadro's number and the mass of the hydrogen atom. The third argues that for high-frequency and low-density blackbody radiation, that is, for the regime in which Wien's distribution law is empirically valid, the entropy of radiation depends on volume in the same way as for an ideal gas, suggesting that radiation of this kind consists of a number of mutually independent quanta of energy $(R/N)\beta\nu$. The fourth considers the consequences of this model of radiation for photoluminescence, the photoelectric effect, and the ionization of gases.

On the assumption of the Light Postulate, the energy of an individual light quantum can be correlated equivalently with the inverse of the wavelength of radiation. Without the Light Postulate, though, the relation between frequency and wavelength is not fixed, and the question can be raised as to which would be fundamental, a fixed correlation of energy with frequency or a fixed correlation with wavelength. The choice has ramifications for predictions concerning the photoelectric effect and ionization when moving sources are used. This question, however, would have been a potential issue for Einstein only if he had considered such correlations prior to abandoning the emission hypothesis.

Curiously enough, the third line of reasoning culminating in the postulation of independent quanta of energy directly proportional to the frequency begins: "Consider a radiation that occupies the volume v. We assume that the observable properties of this radiation are completely determined if the radiation density $\rho(\nu)$ is given for all frequencies" (Einstein 1905a, p. 137). Compare this remark with the passage quoted above from Einstein's June 1912 letter to Ehrenfest.

Query: Is the assumption that the observable properties of radiation are determined by the radiation density tantamount to adoption of the Light Postulate?

This assumption also carries a footnote, which, if the answer to the query is in the affirmative, reflects Einstein's inability to find any direct experimental evidence for or against the Light Postulate: "This assumption is arbitrary. Naturally, one will maintain this simplest assumption as long as experimental results do not force us to abandon it." The purpose of the assumption is to derive the thermodynamic relation

$$\frac{\partial \varphi}{\partial \varrho} = \frac{1}{T},$$

where T is the absolute temperature and φ is the contribution to the entropy per unit volume as a function of ρ and ν. The rest of the argument hangs on this relation. Thus, an affirmative answer to the query would indicate a central role for the Light Postulate in the development of the light quantum hypothesis.

One might also wonder what role, if any, the Principle of Relativity might have played. Implicit in this question is the assumption that Einstein's confidence in the principle did not wane in the meantime.

Query: Did the Principle of Relativity continue as a guiding principle for Einstein throughout his development of the light quantum hypothesis, or did his commitment to it waver at some point or another?

It is tempting to think that Einstein's confidence in the Principle of Relativity did not waver, although I know no way of excluding the possibility that it did. In either case, certain questions concerning the relation between the Principle of Relativity and Einstein's maturing views on light quanta remain.

Query: Would a failure of the light quantum heuristic to conform to the Principle of Relativity yield any testable experimental consequences?

Even if all attempts to measure the absolute motion of the earth by optical means, (that is, using the phenomena of diffraction, reflection, refraction, interference, dispersion, etc.) have been unsuccessful, it is still conceivable that there might exist phenomena involving the production and transformation of light capable of revealing such motion. These might include dependencies in photoelectric or ionization phenomena on the orientation of the experimental apparatus with respect to the earth's motion. Had such considerations occurred to Einstein, it can be surmised that by the time of his drafting the light quantum paper, he had little or no doubt in the validity of the Principle of Relativity. In presenting the mechanical analogy for the interpretation of the entropy formula, he writes: "Concerning the laws, according to which the points under consideration move in the space, nothing shall be assumed except that in regard to this motion no part of the space (*and no direction*) is to be treated with distinction from the others"

(1905a, p. 141; emphasis mine). For the analogy to be coherent, the same should hold for a radiation space, and thus there can be no anisotropy in the laws governing radiation, as would be the case if the Principle of Relativity were not valid. The question naturally arises whether the emerging theory of light quanta is fully consistent with the Principle of Relativity.

Query: What theoretical results concerning light quanta might bear on the consistency of the light quantum hypothesis with the Principle of Relativity?

The most obvious is the relation of energy to frequency. Unless this is invariant, the light quantum hypothesis predicts that certain phenomena concerning the interaction of radiation with matter, for example, the ionization potential of gases, depend on the state of motion of the system.

9. The Electron Problem

It might be claimed that, since the Lorentz and field transformations are needed in order to establish the exact invariance of E/ν, Einstein would not have been in a position to do so until after his discovery of the relativity of simultaneity. This presupposes, however, that he could not have known these transformation equations until after he had laid the kinematical foundations of the special theory. Is there any basis for this supposition?

We have seen evidence suggesting that Einstein had settled on the "time-average" correctness of Maxwell's equations for free space together with the validity of the Principle of Relativity at least several months prior to the discovery of the relativity of simultaneity. Furthermore, it is difficult to imagine that by this time he was not also quite familiar with Lorentz's first-order theorem of corresponding states from the *Versuch*.[27] A natural thing for one to consider then is how the corresponding states theorem can be generalized to an exact theorem and what physical assumptions would be required to ensure null results to all orders. Finding an exact corresponding states theorem is just the problem of producing a set of transformation equations that leave Maxwell's equations formally invariant, which, as we know from the work of Lorentz, is a problem that can be solved without developing a relativistic kinematics.

This motivates our next question.

Query: In what order did Einstein discover (a) the relativistic field transformation equations, (b) the Lorentz transformations, and (c) the relativity of simultaneity?

Even though this question appears unanswerable, given the current corpus of documentary evidence, the various considerations that bear on it are still worth discussing. Perhaps the most significant of these is the way in which Einstein might have approached the so-called electron problem.

Besides the problem of understanding the implications of Planck's formula for blackbody radiation, the other "fundamental crisis" described by Einstein in his *Autobiographical Notes* involved the attempt to replace classical mechanics with the electromagnetic world view as a foundation for all of physics. Fundamental to this program was the conception of the electron as a specially modified region of the aether, and crucial to the success or failure of the program was resolving the question of whether the mass of the electron is entirely electromagnetic in origin or whether it possesses a residual, true "mechanical" mass. In his discussion of the electromagnetic world view in the *Autobiographical Notes*, Einstein also mentions the inherent difficulty on this view of accounting for the stability of the electron—why it does not fly apart as a result of the electrostatic repulsion of its own charge density—without introduction of non-linear equations in place of Maxwell's.

However, with a recognition of the relative character of the field quantities and a corresponding rejection of the aether, the issues of concern for the electromagnetic world view can be seen in an entirely new light. For one thing, there is no longer any question of taking the electron to be a special modification of the aether. Instead, it appears as a self-subsistent, independent structure just as do the light quanta composing the radiation field. For another, it is immediately evident that the very idea of the electromagnetic mass cannot have an objective significance. As was well known, the energy density of the electromagnetic field is given by $E^2 + H^2$. Furthermore, as was frequently discussed at the turn of the century, Lorentz's electrodynamics leads to a violation of the law of action and reaction unless one ascribes a momentum density

$$\frac{1}{c}(\mathbf{E} \times \mathbf{H})$$

to the aether. By integrating these expressions over the entire electromagnetic field of a finite charge distribution (that is, an electron) in motion, one arrives at the ideas of the electromagnetic energy and the electromagnetic momentum associated with the moving charge. For small values of velocity v (in comparison to the velocity of light), these take on the suggestive forms $(1/2)\mu v^2$ and μv respectively, if one sets

$$\mu = \frac{e^2}{6\pi R c^2},$$

where R represents the linear dimensions of the charge distribution and e the total charge. The exact expressions for arbitrary velocities depend on assumptions as to the shape and distribution of the charge. The concepts of longitudinal and transverse electromagnetic mass arise by considering, in the quasi-stationary case, the ratio of the rate of change of electromagnetic momentum G to the accelerations parallel and perpendicular to the direction of motion. These are given by the expressions $d|G|/dv$ and $|G|/v$ respectively, which, though equal for small velocities, diverge for larger ones, the longitudinal electromagnetic mass always being the larger. The mass of the electron can be deemed to be of entirely electromagnetic origin if, in both the longitudinal and transverse directions, the ratio of an applied force to the resulting acceleration equals the electromagnetic mass.

The point here is that the electromagnetic momentum of an electron, and hence the idea of electromagnetic mass, can be traced back to the magnetic field generated by the convection of the electric charge, so that, in effect, as Einstein remarks in the *Autobiographical Notes* (Einstein 1979, p. 35), the magnetic field of a moving charge represents its inertia. But now, if this magnetic field has only a relative existence, and, in fact, always fails to exist in the rest frame of the electron, the notion of electromagnetic mass can be at best an "auxiliary" concept in the same way as the electromotive force exerted on a charge moving through a magnetic field is but an auxiliary concept. Einstein alludes to this fact in the final section of the 1905 relativity paper: "In reference to the *ordinary manner of conception*, we now inquire as to the 'longitudinal' and 'transverse' mass of the moving electron" (Einstein 1905d, pp. 918–919; emphasis mine). Note that all this can be readily grasped without any digression into kinematics.

Once Einstein had rejected the aether and come to view the electric and magnetic fields as frame-dependent quantities, it would have been natural for him to ask whether and in what functional form the "auxiliary" concepts of electromotive momentum and mass can be retrieved as frame effects. That is, if the force acting on an electron moving through an applied electromagnetic field is really just the electric field that exists in the rest frame of the electron, can the usual concepts of longitudinal and transverse mass be derived by transforming back to the "stationary" frame?

Had Einstein considered this problem in the context of an emission theory, he would have faced the preliminary problem of how to conceive of a field in such a context. One alternative is to consider the field at a point of space merely as a representation of the force that would be exerted on a charge were it to occupy that point. (In effect, this is to replace the field with a velocity-dependent potential.) This may not have been Einstein's cup of tea, for the reason that the energy and momentum of a system of charges would no longer be defined at each instant of time, an objection that Einstein later leveled against Ritz's theory (Einstein 1909a, p. 185). Ritz had also suggested, as an aid to the imagination, the mediation of electromagnetic interactions via fictive particles. Einstein might possibly have considered a picture such as this, only with real particles (light quanta?) instead of fictive ones. How far he might have gotten is anyone's guess. Keep in mind, though, that Ritz found the properly electrodynamical problems more tractable than the optical problems from the point of view of an emission theory and succeeded in producing a definite theory of the dynamics of the electron for which he claimed quantitative agreement with Lorentz's 1904 theory (and hence with special relativity).

However, once one has come to the conclusion that the Maxwell-Hertz equations are correct at least for "time-average" average values, the electric and magnetic fields can then be seen to represent an approximate description of actual processes going on in free space. This forces the question of how these field quantities transform quantitatively from one frame of reference to another. According to Lorentz's expression for the electromotive force, the electric field must transform, at least to first-order in v/c as

$$\mathbf{E}' = \mathbf{E} + \frac{1}{c}(\mathbf{v} \times \mathbf{H}).$$

The analogy with "magneto-motive" forces suggests that the magnetic field transforms to first-order according to

$$\mathbf{H}' = \mathbf{H} - \frac{1}{c}(\mathbf{v} \times \mathbf{E}).$$

Both expressions are ubiquitous in the contemporary literature, including the *Versuch*. That these expressions can be valid only to first-order is clear from a "reciprocity requirement" to the effect that the result of transforming the fields from one frame of reference to another and then back again must yield the original fields. Thus, one is motivated to cast about for the correction terms needed to satisfy the reciprocity requirement.

Clark Glymour has suggested a strategy here that rigorously yields the relativistic field transformations. Assuming the velocity v is in the positive x-direction, one surmises that the transformations should have the form

$$E'_x = \varphi_x(v/c)E_x$$

$$E'_y = \varphi_y(v/c)(E_y - v/cH_z)$$

$$E'_z = \varphi_z(v/c)(E_z + v/cH_y)$$

and

$$H'_x = \psi_x(v/c)H_x$$

$$H'_y = \psi_y(v/c)(H_y + v/cE_z)$$

$$H'_z = \psi_z(v/c)(H_z - v/cE_y).$$

The problem is to solve for the factors φ_i and ψ_i. Symmetry between the y- and z-directions together with the reciprocity requirement demands first that

$$\varphi_x = 1 = \psi_x$$

and second that

$$\varphi_y = \varphi_z = (1 - v^2/c^2)^{-1/2} = \psi_y = \psi_z.$$

This yields the exact relativistic field transformation equations.

By substituting these into the Maxwell-Hertz equations for free space, the Lorentz transformations can be obtained in a fashion that is just the inverse of the derivation of the field transformations from the Lorentz transformations in section 6 of the relativity paper. Although the use of such velocity-dependent multiplicative factors is reminiscent of the techniques used by Einstein in the 1905 relativity paper, the claim here is not that Einstein indeed derived the field transformations in this fashion, or even that he obtained the field transformations prior to the Lorentz transformations, but only that it is solidly within the realm of historical plausibility for him to have been in possession of both sets of transformations without having first pondered the problem of distant simultaneity. And given these transformations, the various formulas that appear in section 10 on the dynamics of the slowly accelerated electron, as well as those in sections 7 and 8, are readily obtainable.

The physical meaning of what has been done, however, would be enigmatic. Given the relative character of the field quantities and the assumption of the "time-average" correctness of the Maxwell-Hertz equations, it follows that the phenomena behave *as though* the Lorentz transformations rather than the Galilean transformations express the relation between coordinates in different uniformly moving frames. That this should be so makes some preliminary sense in that it would account for the apparent adequacy of the Lorentz-Fitzgerald contraction hypothesis (discussed in Wien 1898, as well as in Lorentz's *Versuch*). On the other hand, *why* it should be so is still less than obvious. This would have been sufficient to put Einstein in a state of severe perplexity, precipitating the events of late May 1905.

10. Point and Counterpoint

This suggestion will, no doubt, draw a number of objections. Let me rehearse what I take to be the obvious ones and attempt to disarm them.

First, it might be objected that had Einstein been in possession of the Lorentz transformations prior to his discovery of the relativity of simultaneity, then he would have already known that the Principle of Relativity and the Light Postulate are consistent with one another, contrary to his later characterizations of the state of crisis that lead to that discovery. There is, however, a subtlety in the notion of consistency that should not be overlooked. One can know that certain equations are mathematically consistent, that is, that they do not lead to a formal contradiction, without knowing whether they are *physically* consistent, that is, whether they admit a physical interpretation that does not lead to consequences in conflict with experience or other secure physical principles. And without an understanding of the physical basis for the Lorentz transformations, there is no guarantee as to the physical consistency of the Principle of Relativity and the Light Postulate.

Another objection might be that, given the Lorentz transformations, the relativity of simultaneity, and thus the key to their physical interpretation, is all too obvious. In reply, the relativity of simultaneity is just as "obvious" from the first-order Lorentz transformations, and Einstein was no doubt quite familiar with these from the *Versuch*. In truth, the rejection of absolute simultaneity is not at all an obvious step to take, but is something that requires a careful and searching epistemological analysis in order to break a deep-seated habit of thought. Nor are the consequences to be taken lightly. For as a result the objectivity of the very notions of past, present, and future is undermined. Upon the death of his life-long friend Michele Besso in 1955, Einstein consoled Besso's son and sister: "Now he has preceded me a bit even in departing from this strange world. This means nothing. For us believing physicists, the divide between past, present, and future has only the significance of an illusion, albeit a stubborn one" (Speziali 1972, p. 538).

On the other side, there is a consideration that suggests that the sort of reflections that enter into the discovery of the conventional character of simultaneity cannot get on the right track without an awareness of the Lorentz transformations as a live alternative to the Galilean transformations. If one is thinking about the operational meaning of the basic kinematical concepts, the obvious definition of clock synchronization that suggests itself is to take a master clock and transport it around to set all your other clocks. From the point of view of special relativity, this is problematic because of time dilation. But what basis is there for suspecting time dilation other than the Lorentz transformations? Note that this is something that cannot be surmised from Lorentz's first-order local time.

11. The "Fizeau Problem"

In surveying the logical structure of the 1905 relativity paper, it was possible to identify two major problem areas of interest to Einstein—the electron problem and the radiation-structure problem—both of which received prominent attention in the *Autobiographical Notes*. There is perhaps yet a third problem area that can be distilled from the structure of the paper.

The one experimental result most often cited by Einstein in connection with the genesis of special relativity is that of Fizeau's experiment. Oddly enough, this experiment is nowhere mentioned in the 1905 paper. As I hinted earlier, one can nevertheless construe section 9 as implicitly providing an explanation of the result since it is a by-product of the demonstration that Lorentz's electron theory conforms with the Principle of Relativity. Crucial to this derivation is the relativistic velocity addition rule obtained in section 5.

Now it may be that Fizeau's experiment played a significant role only at earlier junctures in the development of the theory. We have seen evidence suggesting that it factored crucially into the rejection of a movable aether. It may also have entered into Einstein's reasons for rejecting the emission hypothesis. In an unpublished essay from 1912 on the foundations of electrodynamics,[28] Einstein explicitly considered the difficulties that arise from attempting to account for the experiment if the velocity of light depends on the motion of the source, although it is unclear to what extent the dialectic presented there might reflect a stage of reasoning prior to 1905.

There is, however, a consideration that suggests how Fizeau's experiment might have figured in yet another way. For a transparent medium of index of refraction n, the velocity of propagation of light through the medium is c/n. Fizeau's experiment establishes that if the medium is set in motion with a velocity v, then the velocity of light through the medium as measured in the stationary frame is

$$V_{\text{stat}} = c/n + v(1 - 1/n^2).$$

Using the classical velocity addition rule, it immediately follows from this that the velocity of light relative to the frame of the moving medium is

$$V_{\text{mov}} = (c/n - v/n^2),$$

rather than c/n. One might conclude, as did Abraham, that this establishes the invalidity of the "Axiom of Relative Motion" for the optics of moving bodies (Abraham 1904c, p. 435). In retrospect, we can see that this conclusion is a bit hasty. All that follows is that the classical velocity addition rule and the Principle of Relativity cannot both be valid. What is significant, though, is that Fizeau's result alone apparently suffices to establish this.

I draw from this the conclusion that there are a number of important questions that have yet to be formulated. No doubt there are numerous other considerations that would suggest the same.

Acknowledgments. An earlier version of this paper was presented at the Conference on the Early Einstein, Osgood Hill, October 1990. I would like to single out John Stachel and John Norton for encouragement and advice, David Cassidy and Robert Schulmann for their assistance and hospitality, and Gerald Holton, Don Howard, Martin Klein, and Jürgen Renn for helpful conversations. I would also like to thank Don Howard for his editorial efforts and advice. I am grateful to the Hebrew University of Jerusalem, which holds the copyright, for permission to quote from various unpublished letters and manuscripts in the Einstein Archive, and to the National Endowment for the Humanities for making travel to that collection possible.

The present essay grew out of a project undertaken with John Earman and Clark Glymour during the years 1979–1982, which resulted in many unpublished pages on the problem of the construction of the special theory (see, however, Earman et. al. 1983). Many of the ideas suggested above stem from that collaboration. Although my debt to Earman and Glymour is great, I do not know the extent to which they would approve of what has been presented here.

NOTES

[1] Stark had suggested that Einstein write a book on "das Relativitätsprinzip." Einstein replied: "Unfortunately it is impossible for me to write this book, because it is impossible for me to find the time for it. Each day 8 hours of strenuous work at the patent office, and in addition much correspondence and studying" (Einstein to Stark, 14 December 1908, in Hermann 1966, p. 277).

[2] The last paper he had written prior to this, Einstein 1904, is dated 27 March 1904.

[3] See especially Einstein to Besso, 17 March 1903, in Speziali 1972, pp. 13–14.

[4] That paper begins: "As great as the achievements of the kinetic theory of heat have been in the area of the theory of gases, nevertheless mechanics has not hitherto been in a position to provide a satisfactory foundation for the general theory of heat, because it has not yet succeeded in deriving the laws of thermal equilibrium and the second law of thermodynamics from the equations of mechanics and the probability calculus alone, although Maxwell's and Boltzmann's theories have already come close to this goal. The purpose of the following considerations is to fill this gap" (Einstein 1902b, p. 417).

[5] This was probably Besso, whom Einstein thanks at the conclusion of the 1905 relativity paper.

[6] Note also Henri Poincaré's remark near the very conclusion of *La science et l'hypothèse*: "Imagine two charged conductors with the same velocity of translation. They are relatively at rest. However, each of them being equivalent to a current of convection, they ought to attract one another, and by measuring this attraction we could measure their absolute velocity. 'No!' replied the partisans of Lorentz. 'What we could measure in that way is not their absolute velocity, but their relative velocity *with respect to the ether*, so that the principle of relativity is safe'" (Poincaré 1902, pp. 243–244). Einstein had read *La science et l'hypothèse* sometime around 1903 as part of his discussion group, the "Olympia Academy"

(Solovine 1956, p. Viii).

[7] See Einstein and Infeld 1938, pp. 164–177 for a discussion.

[8] Occasionally, stellar aberration is mentioned as well. See Einstein 1922b, p. 27; Shankland 1963, p. 48; "Grundgedanken und Methoden der Relativitätstheorie, in ihrer Entwicklung dargestellt," EA 2-022-2; "Das Wesen der Relativitätstheorie," EA 2-115-2; and draft page for the *Autobiographical Notes*, EA 2-0022-2.

[9] Reiser 1930, p. 49; Frank 1949, p. 38. See Holton 1967–8 for further details.

[10] The editors of Einstein 1987 identify this as H.F. Weber, Einstein's physics supervisor.

[11] This may be a reference to Drude 1900, which contains a discussion of Lorentz's theory.

[12] From a draft page probably for the *Autobiographical Notes*, EA 2-022-2.

[13] For example, Paul Ehrenfest, Walter Ritz, Richard C. Tolman, J. Kunz, D.F. Comstock, and O.M. Stewart; see Pauli 1921, pp. 5–8 for references.

[14] "Die in dem Artikel ausgesprochene Hypothese, dass das Licht im leeren Raum nicht gegen das Coordinatensystem sondern gegenüber der Lichtquelle die konstante Geschwindigkeit *c* habe, ist zum ersten Mal von dem schweizerischen Physiker W. Ritz ausführlich diskutiert und von mir selbst vor Aufstellung der speziellen Relativitätstheorie ernsthaft in Betracht gezogen worden." See also the letter of Einstein to Ehrenfest, 25 April 1912, EA 9-324 and the draft of a letter to A.P. Rippenbein, *circa* 1952, EA 20-046.

[15] "Ausserdem verlangt diese Theorie, dass überall und in jeder bestimmten Richtung Lichtwellen verschiedener Fortpflanzungsgeschwindigkeit möglich sein sollen. Es dürfte unmöglich sein, eine irgendwie vernüftige elektromagnetische Theorie aufzustellen, die solcher leistet. Dies ist der hauptsächliche Grunde, aus dem ich schon vor der Aufstellung der speziellen Relativitätstheorie diesem an sich denkbaren Ausweg verworfen habe."

[16] "Grundgedanken und Methoden der Relativitätstheorie, in ihrer Entwicklung dargestellt." Translation by Gerald Holton; quoted from *The New York Times*, 28 March 1972, p. 32.

[17] Keep in mind that the term "electron" as used at the time is not coextensive with our current usage of the term, but is synonymous with "fundamental charged particle." One finds frequent references from the period to the "positive electron," and it is certainly not the positron that was in question. See, for example, Lorentz 1909, p. 8 and Hilbert 1926, p. 185.

[18] "Wenn man Deine Kollektion der Bestätigungen der speziell. Rel. Theorie durchgeht, so meint man, die Maxwell'sche Theorie sei zum Greifen sicher. Aber 1905 wusste ich schon sicher, dass sie zu falschen Schwankungen des Strahlungsdruckes führt und damit zu einer unrichtigen Brown'schen Bewegung eines Spiegels in einem Planck'schen Strahlungs-Hohlraum."

[19] Boltzmann's *Gastheorie* (Boltzmann 1896, 1898) had made a deep impression on Einstein when he read it in 1900; see Einstein to Marić, 19 September 1900, in Einstein 1987, Doc. 75, p. 260. Einstein's three papers on

statistical thermodynamics can be seen as an attempt to generalize the statistical theory of heat beyond the domain of gases so as to be applicable to liquids, solids, and, as is suggested here, to radiation, as well.

[20] Although this consideration might have crystallized only after 1905, such musings might very well have occurred to Einstein as early as April of 1901. In a pair of letters to Marić from that month Einstein mentions his negative reaction to what appears to be Planck's paper, "On Irreversible Radiation Processes" (Planck 1900). See Einstein 1987, Doc. 96, p. 284 and Doc. 97, p. 286.

[21] The editors of volume 2 of Einstein's *Collected Papers* note: "This agreement confirms the applicability of statistical concepts to radiation, and may have suggested to [Einstein] the possibility that radiation could be treated as a system with a finite number of degrees of freedom, a possibility he raised at the outset of his first paper on the quantum hypothesis" (Einstein 1989, p. 138).

[22] "Ich verwarf diese Hypothese damals, weil sie zu grossen theoretischen Schwierigkeiten führt (z.B. Erklärung der schattenbildung durch einer Schirm, der relativ zur Lichtquelle bewegt ist)."

[23] "Man kann für die Hypothese der Unabhängigkeit der Lichtgeschwindigkeit vom Bewegungszustande der Lichtquelle wohl nur deren Einfachheit und leichte Durchfürbarkeit anführen. Sobald man diese Hypothese aufgibt, muss man schon zur Erklärung der Schattenbildung die hässliche Voraussetzung einführen, dass das von einem Resonator emittierte Licht von der Art der Erregung abhänge (Erregung durch 'bewegte' Strahlung oder Erregung anderer Art)."

[24] Einstein to C.O. Hines, February 1952: "If an appropriately accelerated light source emits light in one direction, (e.g., in the direction of acceleration), then the planes of equal phase move with different velocities, and thus one can arrange it that all the surfaces of equal phase come together at a given location, so that the wave length there becomes infinitely small. From there on the light reverses itself, so that the rear part overtakes the front." ["Wenn eine passend beschleunigte Lichtquelle Licht in einer Richtung, (z.B. der Richtung der Beschleunigung) aussendet, so bewegen sich die Ebenen gleicher Phase mit verschiedener Geschwindigkeit, und man kann es so einrichten, dass alle die Flächen gleicher Phase zusammenfallen an einem gewissen Orte, sodass die Wellenlänge dort unendlich klein wird. Darüber hinaus wird sich das Licht so umdrehen, dass der hintere Teil den vorderen überholt"] (EA 12-251).

According to Robert Shankland: "He gave up [the emission hypothesis] because he could think of no form of differential equation which could have solutions representing waves whose velocity depended on the motion of the source. In this case, the emission theory would lead to phase relations such that the propagated light would be all badly 'mixed up' and might even 'back up on itself'" (Shankland 1963).

[25] Einstein to C.O. Hines, February 1952: "Further, an ever so thin permeable film would modify the velocity of 'moving' light by a finite amount, so that the interference . . . would give rise to entirely unbelievable phenomena." ["Ferner wird eine noch so dünne durchlässige Schicht die Lichtgeschwindigkeit von 'bewegtem'

Licht um einen endlichen Betrag verändern, sodass die Interferenz . . . zu ganz unglaubhaften Phaenomenen Veranlassung gäbe"] (EA 12-251).

Einstein to Ehrenfest, April 25, 1912: "In case 2) [the case in which 'moving' light transforms into 'stationary' light] the following must happen: Let S_1 and S_2 be two coherent rays, which come from a moving star, one of which passes through a membrane *B*. The phase difference of the two rays at *E* depends then on the distance *B-E* in quantities of the first order, since the wavelength is different before and after *B*" (EA 9-324; Einstein 1993, Doc. 384, p. 450).

[26] Einstein to C.O. Hines, February 1952: "But the strongest argument seemed to me: If in general there is no fixed light velocity, then why should it be that all light which is emitted by 'stationary' bodies has a velocity *completely independent of the color*? This seemed absurd to me." ["Das stärkste Argument aber erschien mir: Wenn es nicht überhaupt eine feste Lichtgeschwindigkeit gibt, warum soll es dann so sein, dass alles Licht, was von 'ruhenden' Körpern ausgeht, ein *von der Farbe völlig unabhängige* Geschwindigkeit hat? Dies erschien mir absurd"] (EA 12-251).

[27] For a discussion of how Lorentz employed this theorem to account for the absence of first-order effects of the earth's motion, see Rynasiewicz 1988.

[28] "Manuscript on the Special Theory of Relativity," in Einstein 1995, Doc. 1, pp. 9–101.

References

Abraham, Max (1902a). "Dynamik des Elektrons." *Königliche Gesellschaft der Wissenschaften zu Göttingen. Mathematisch-physikalische Klasse. Nachrichten*: 20–41.

— (1902b). "Prinzipien der Dynamik des Elektrons." *Physikalische Zeitschrift* 4: 57–62.

— (1903). "Prinzipien der Dynamik des Elektrons." *Annalen der Physik* 10:105–179.

— (1904a). "Der Lichtdruck auf einen bewegten Spiegel und das Gesetz der schwarzen Strahlung." In Meyer 1904, pp. 85–93.

— (1904b). "Zur Theorie der Strahlung and des Strahlungsdruckes." *Annalen der Physik* 14: 236–287.

— (1904c). *Theorie der Elektrizität*. Vol. 1, August Föppl, *Einführung in die Maxwell'sche Theorie der Elektrizität*, 2nd rev. ed. Max Abraham, ed. Leipzig: B.G. Teubner.

Boltzmann, Ludwig (1884). "Ableitung des Stefan'schen Gesetzes, betreffend die Abhängigkeit der Wärmstrahlung von der Temperatur aus der electromagnetischen Lichttheorie." *Annalen der Physik und Chemie* 22: 291–294.

— (1896). *Vorlesungen über Gastheorie*. Part 1, *Theorie der Gase mit einatomigen Molekülen, deren Dimensionen gegen die mittlere Weglänge verschwinden*. Leipzig: Johann Ambrosius Barth.

— (1898). *Vorlesungen über Gastheorie*. Part 2, *Theorie Van der Waals'; Gase mit zusammengesetzten Molekülen; Gasdissociation; Schlussbemerkungen*. Leipzig: Johann Ambrosius Barth.

Drude, Paul (1900). *Lehrbuch der Optik*. Leipzig: S. Hirzel.

Earman, John; Glymour, Clark and Rynasiewicz, Robert (1983). "On Writing the History of Special Relativity." *PSA 1982*, vol. 2. Peter D. Asquith and Thomas Nickles, eds. East Lansing: Philosophy of Science Association, pp. 403–416.

Einstein, Albert (1901). "Folgerungen aus den Capillaritätserscheinungen." *Annalen der Physik* 4: 513–523.

— (1902a). "Über die thermodynamische Theorie der Potentialdifferenz zwischen Metallen und vollständig dissociirten Lösungen ihrer Salze und über eine elektrische Methode zur Erforschung der Molecularkräfte." *Annalen der Physik* 8: 798–814.

— (1902b). "Kinetische Theorie des Wärmegleichgewichtes und des zweiten Hauptsatzes der Thermodynamik." *Annalen der Physik* 9: 417–433.

— (1903). "Eine Theorie der Grundlagen der Thermodynamik." *Annalen der Physik* 11: 170–187.

— (1904). "Zur allgemeinen molekularen Theorie der Wärme." *Annalen der Physik* 14: 354–362.

— (1905a). "Über einen die Erzeugung und Vervandlung des Lichtes betreffenden heuristischen Gesichtspunkt." *Annalen der Physik* 17: 132–148.

— (1905b). *Eine neue Bestimmung der Moleküldimensionen*. Bern: K. J. Wyss. A slightly revised version published as "Eine neue Bestimmung der Moleküldimensionen." *Annalen der Physik* 19 (1906): 289–305.

— (1905c). "Über die von der molekularkinetischen Theorie der Wärme geforderte Bewegung von in ruhenden Flüssigkeiten suspendierten Teilchen." *Annalen der Physik* 17: 549–560.

— (1905d). "Zur Elektrodynamik Bewegter Körper." *Annalen der Physik* 17: 891–921.

— (1909a). "Zum gegenwärtigen Stand des Strahlungsproblems." *Physikalische Zeitschrift* 10: 185–193.

— (1909b). "Über die Entwickelung unserer Anschauungen über das Wesen und die Konstitution der Strahlung." *Deutsche Physikalische Gesellschaft. Verhandlungen* 7: 482–500.

— (1917). *Über die spezielle und die allgemeine Relativitätstheorie. (Gemeinverständlich)*. Braunschweig: Friedrich Vieweg und Sohn. Page numbers and quotations from the English translation: *Relativity: The Special and the General Theory. A Clear Explanation that Anyone Can Understand.* Robert W. Lawson, trans. New York: Bonanza Books, 1961.

— (1922a). "How I Created the Theory of Relativity." Address at Kyoto University, 14 December 1922. Yoshimasa A. Ono, trans. *Physics Today*, August 1982, 45–47.

— (1922b). *The Theory of Relativity*. Princeton: Princeton University Press. Page numbers from the 5th ed., 1953.

— (1956). *Lettres à Maurice Solovine*. Maurice Solovine, ed. and trans. Paris: Gauthier-Villars.

— (1979). *Autobiographical Notes: A Centennial Edition*. Paul Arthur Schilpp, trans. and ed. La Salle, Illinois: Open Court. Corrected version of "Autobiographisches—Autobiographical Notes." In *Albert Einstein: Philosopher-Scientist*. Paul Arthur Schilpp, ed. Evanstan, Illinois: The Library of Living Philosophers, 1949, pp. 1–94.

— (1987). *The Collected Papers of Albert Einstein*. Vol. 1, *The Early Years, 1879–1902*. John Stachel, et al., eds. Princeton: Princeton University Press.

— (1989). *The Collected Papers of Albert Einstein*. Vol. 2, *The Swiss Years: Writings, 1900–1909*. John Stachel, et al., eds. Princeton: Princeton University Press.

— (1993). *The Collected Papers of Albert Einstein*. Vol. 5, *The Swiss Years: Correspondence, 1902–1914*. Martin J. Klein, et al., eds. Princeton: Princeton University Press.

— (1995). *The Collected Papers of Albert Einstein*. Vol. 4, *The Swiss Years: Writings, 1912–1914*. Martin J. Klein, et al., eds. Princeton: Princeton University Press.

Einstein, Albert and Infeld, Leopold (1938). *The Evolution of Physics*. New York: Simon and Schuster.

Föppl, August (1894). *Einführung in die Maxwell'sche Theorie der Elektricität*. Leipzig: B.G. Teubner.

Frank, Philipp (1949). *Einstein: Sein Leben un seine Zeit*. Munich: Paul List.

Hasenöhrl, Fritz (1904). "Zur Theorie der Strahlung in bewegten Körpern." *Annalen der Physik* 15: 344–370.

Hermann, Armin (1966). "Albert Einstein und Johannes Stark: Briefwechsel und Verhältnis der beiden Nobelpreisträger." *Sudhoffs Archiv. Vierteljahrsschrift für Geschichte der Medizin und der Naturwissenschaften, der Pharmazie und der Mathematik* 50: 267–285.

Hilbert, David (1926). "Über das Unendliche." *Mathematische Annalen* 95: 161–190. Page numbers from the English translation: "On the Infinite." Erna Putnam and Gerald J. Massey, trans. In *Philosophy of Mathematics*, 2nd ed. Paul Benacerraf and Hilary Putnam, eds. Cambridge: Cambridge University Press, 1983, pp. 183–201.

Holton, Gerald (1967–8). "Influences on Einstein's Early Work." *The American Scholar* 37: 59–79. Reprinted in Holton 1973, pp. 197–217.

— (1969). "Einstein, Michelson, and the 'Crucial' Experiment." *Isis* 60: 133–197. Page numbers and quotations taken from the reprinting in Holton 1973, pp. 261–352.

— (1973). *Thematic Origins of Scientific Thought: Kepler to Einstein*. Cambridge, Massachusetts: Harvard University Press.

— (1980). "Einstein's Scientific Program: The Formative Years." In *Some Strangeness in the Proportion: A Centennial Symposium to Celebrate the*

Achievements of Albert Einstein. Harry Woolf, ed. Reading, Massachusetts: Addison-Wesley, pp. 49–65.

Kretschmann, Erich (1917). "Über den physikalischen Sinn der Relativitätspostulate, A. Einsteins neue und seine ursprüngliche Relativitätstheorie." *Annalen der Physik* 53: 575–614.

Lorentz, Hendrik Antoon (1895). *Versuch einer Theorie der electrischen und optischen Erscheinungen in bewegten Körpern*. Leiden: E.J. Brill.

— (1904). "Electromagnetic Phenomena in a System Moving with Any Velocity Smaller Than That of Light." *Koninklijke Akademie van Wetenschappen te Amsterdam. Section of Sciences. Proceedings* 6: 809–831. Reprinted in *The Principle of Relativity*. W. Perrett and G.B. Jeffery, trans. London: Methuen, 1923; reprint New York: Dover, 1952, pp. 11–34.

— (1909). *The Theory of Electrons and Its Applications to the Phenomena of Light and Radiant Heat*. Leipzig: B.G. Teubner; 2nd ed., 1915, reprint New York: Dover, 1952.

Meyer, Stefan, ed. (1904). *Festschrift. Ludwig Boltzmann gewidmet zum sechzigsten Geburtstage 20 Februar 1904*. Leipzig: Johann Ambrosius Barth.

Pauli, Wolfgang (1921). "Relativitätstheorie." In *Encyklopädie der mathematischen Wissenschaften, mit Einschluss ihrer Anwendungen*. Vol. 5, *Physik*, part 2. Arnold Sommerfeld, ed. Leipzig: B.G. Teubner, 1904–1922, pp. 539–775. [Issued November 15, 1921.] English translation: *Theory of Relativity*. With supplementary notes by the author. G. Field, trans. London: Pergamon, 1958; reprint New York: Dover, 1981.

Planck, Max (1900). "Über irreversible Strahlungsvorgänge." *Annalen der Physik* 1: 69–122.

Poincaré, Henri (1902). *La science et l'hypothèse*. Paris: E. Flammarion. Page numbers and quotations from the English translation: *Science and Hypothesis*. New York: Dover, 1952.

Reiser, Anton [Rudolph Kayser] (1930). *Albert Einstein: A Biographical Portrait*. New York: Albert and Charles Boni.

Rynasiewicz, Robert (1988). "Lorentz's Local Time and the Theorem of Corresponding States." *PSA 1988*, vol. 1. Arthur Fine and Jarrett Leplin, eds. East Lansing: Philosophy of Science Association, pp. 67–74.

Seelig, Carl (1960). *Albert Einstein: Leben und Werk eines Genies unserer Zeit*. Zürich: Europa Verlag.

Shankland, Robert (1963). "Conversations with Albert Einstein." *American Journal of Physics* 31: 47–57.

Solovine, Maurice (1956). "Introduction." In Einstein 1956.

Speziali, Pierre, ed. (1972). *Albert Einstein—Michele Besso. Correspondance, 1903– 1955*. Paris: Hermann.

Stachel, John (1982). "Einstein and Michelson: The Context of Discovery and the Context of Justification." *Astronomische Nachrichten* 303: 47–53.

— (1987). "Einstein and Ether Drift Experiments." *Physics Today*, May, 45–47.

Wertheimer, Max (1945). *Productive Thinking*. New York and London: Harper & Brothers.

Wien, Wilhelm (1893). "Eine neue Beziehung der Strahlung schwarzer Körper zum zweiten Hauptsatz der Wärmetheorie." *Königlich Preussische Akademie der Wissenschaften* (Berlin). *Sitzungsberichte*: 55–62.

— (1894). "Temperatur und Entropie der Strahlung." *Annalen der Physik und Chemie* 52: 132–165.

— (1898). "Über die Fragen, welche die translatorische Bewegung des Lichtäthers betreffen." *Annalen der Physik und Chemie* 65, no. 3 (Beilage), pp. i–xviii.

Physical Approximations and Stochastic Processes in Einstein's 1905 Paper on Brownian Motion

Sahotra Sarkar

1. Approximations

This paper has three purposes: (i) to point out how much of the modern theory of stochastic processes was foreshadowed in the second of Einstein's three great papers from 1905, the one on Brownian motion (Einstein 1905c);[1] (ii) to explore how the critical derivation in that paper consists of an intriguing series of "physical approximations"; and (iii) to explore some of the mathematical contexts—mainly outside physics—in which that paper appeared, and those that it helped create. The first two points are connected. As A.J. Leggett has recently emphasized, derivations in scientific papers rarely consist of what a mathematician would recognize as a "derivation." What they amount to is

> a hybrid process, in which some steps are indeed mathematically rigorous, while others—the so-called "physical approximations"—are actually not approximations in the usual sense at all, but rather are more or less intelligent guesses, guided perhaps by experience with related systems. . . . Mathematical and physical arguments are intimately intertwined, often without explicit comment. (Leggett 1987, p. 116)

If more than one "physical approximation" is made during the construction of a model, Leggett's scheme only requires them to be "mutually compatible."[2]

What emerges from Leggett's discussion is a view of physical theories rather different from the ones usually doled out by philosophers of science, whether they be old-fashioned logical empiricists or their allegedly more sophisticated descendants.[3] Not only are putatively universal theories replaced by clusters of models, but these models also cannot be arranged

Einstein Studies, vol. 8: Einstein: The Formative Years, pp. 203–229.

in a "logical" hierarchy on the basis of "explanatory strength"—the rules and the concepts that the different models employ are "autonomous" (to use Leggett's term).[4] The only generic relation between models is that of "physical approximation." This, as Leggett says, is often a "guess" about the physics of the situation, a physical hypothesis usually of a very special sort: It only stipulates how to describe a situation—which parameters are relevant, and which are not.

A conventionally obvious question is whether such a hypothesis is true. The answer is rarely straightforward. Leave aside the (basically correct) point that every model is potentially subject to eventual correction, and that it is notoriously difficult to move from the "correctness" of a model to its "truth" (in the philosophers' semantic sense). The "incorrectness" of an approximation does not necessarily signal the failure of some physical principle; rather, it may be the description that is to blame. Beyond these points, it is often the case that two different sets of approximations may yield the same experimental consequences and, moreover, their autonomy suggests their possible non-convergence. What the autonomy of different models (embodying different approximations) does, moreover, is to suggest a rather different and interesting question about a model: whether an approximation is fecund, that is, how well it serves as a resource for further inquiry.

Few papers exhibit the last trait better than Einstein's 1905 paper on Brownian motion, and it does so with several additional interesting twists. It should be regarded as a masterpiece of formal heuristics, that is, the use of intuitive formal steps.[5] The crucial derivation in it consists of a set of well-motivated (and explicit) physical approximations. Several of these point to models that have found natural use in other contexts, in astrophysics, population biology, information theory, and elsewhere. The reason why these models found employment elsewhere is that the research programs that extended and clarified the methods of Einstein's paper—and other similar work from roughly the same period—led to the systematic development of a new type of *mathematical* theory, that of stochastic processes. This development was helped by the almost simultaneous introduction of a formal random walk problem by Karl Pearson (1905a). The result was the codification of "Brownian motion" as a standard stochastic process, but not as a unique one—there are a variety of stochastic models that are all called "Brownian motion"—and it has become progressively less clear how each of them is to be related (even by stipulation) either to Einstein's original model or to the physical process to which he was apparently referring, that is, Brownian motion.[6] In the case of Brownian motion, mathematics and physics are perhaps even more intimately related than Leggett indicates.[7]

This paper will take up only that part of this story that is directly relevant to Einstein's original paper.[8] Even that will require some careful attention to differences between various types of approximation. A full taxonomy of these is also beyond the scope of this paper. However, it is easy enough to amplify—and extend—Leggett's discussion to make six sets of crude distinctions:

(i) Approximations may be explicit or implicit. There are at least two standard strategies of implicit approximation: (a) to invoke a customary procedure that implicitly makes an approximation (ignore gravitational terms while constructing a Hamiltonian, leave out three-particle collisions, etc.); and (b) to invoke a model or formula that makes such an approximation (see Einstein's use of Stokes's formula below).[9]

(ii) Approximations may be corrigible, incorrigible in practice, or incorrigible in principle. Corrigibility is *not* to be construed absolutely. Rather, all that is required is the knowledge of some procedure for decreasing the effects of an approximation. Usually, this involves a procedure for introducing corrections. An example (discussed below) is the attempt to correct Stokes's formula. There is no general procedure for determining when an incorrigibility in practice reflects an incorrigibility in principle. In a particular context, however, it is usually possible to make this judgement.[10]

(iii) The maximal effects of an approximation may be estimable, not estimable in practice, or not estimable in principle. The question of estimability is different from that of corrigibility, because even if the effect of an approximation can be estimated, say up to an upper bound, it need not be removable. It will be pointed out below that (if Einstein's model of Brownian motion is taken as the standard) Langevin's model involves an approximation whose effects are not estimable (at least in practice).

(iv) Approximations may involve: (a) mathematically justified procedures (such as taking limits); (b) physically justified procedures for transitions from one formal statement to another; or (c) neither. "Physical justification" is as strong a condition as mathematical consistency. Justification must come from *prior* physical commitments, that is, not through the implicit introduction of new physics into a derivation.[11] As an example, consider the (classical) thermodynamic limit: $n \to \infty$, $V \to \infty$, with $n/V =$ constant, where n is the number of particles, and V the volume of the system. This limit is physically justified since it captures the classical idea of the increase of the size of a system. It is mathematically justified only if it exists, which is rare (a particularly problematic example will be discussed below).

(v) Approximations may be context-dependent or context-independent. Consider the thermodynamic limit again. From a mathematically rigorous point of view, for most classical systems (such as a system of particles with Coulomb interactions), it is easy to show that this limit does not exist.[12] There is one well-known supposed exception. This is a system of particles that have a specified short-range interaction (of say, a range $\leq R/2$, where R is a characteristic length; see Thompson 1972). Then it can be shown that the limit exists by constructing a model with $n \rightarrow \infty$, $V \rightarrow \infty$, with $n/V =$ constant, by putting the particles in cells separated by walls of thickness R. This is still an approximate procedure, and one of a rather unsatisfactory sort, since nothing in the assumed physics suggests that a macroscopic limit should be constructed in this cellular way. Moreover, this approximation, introduced precisely to accommodate the particular interaction at hand, is about as context-dependent as an approximation can be.

(vi) Approximations may involve counterfactual assumptions, or not. Throughout this paper, the term "counterfactual" is to be construed simply as referring to assumptions not permitted by the physics that is explicitly assumed. Counterfactual assumptions are endemic, though the extent to which they involve "serious" violations of physics is often hard to gauge.[13] Arguably, classical physics does not even allow truly rigid bodies. More clearly, any limiting procedure that allows $V \rightarrow \infty$ or $n \rightarrow \infty$ involves a counterfactual assumption in some sense. Assuming that $c \rightarrow \infty$, or $h \rightarrow 0$, the standard procedures for the recovery of classical physics from special relativity and quantum mechanics, respectively, are also counterfactual. The first of these is not particularly problematic, partly because it is mathematically better-defined than the second; the classical limit of quantum mechanics, however, remains one of the least understood limits in physics.

If an approximation, preferably explicit, is corrigible, its effects estimable, both mathematically and physically justified, and context-independent, and if it involves no counterfactual assumption, then it is presumably philosophically unproblematic. As Leggett notes, philosophers of physics often assume that derivations involve no approximation, or only approximations of this sort, in which case they can potentially be removed to recover the mathematical cleanliness that these philosophers deify. The trouble is that physics rarely proceeds according to such strictures.[14] More than that, the approximations that violate many of these strictures, especially those approximations that are physically or mathematically unjustified (according to prior assumptions), potentially open the way to new developments in physics and mathematics. Dirac's δ-function is perhaps the best-known of such developments—more will be considered below. The ability to intuit which violations are "reasonable," and which are simply errors, is no doubt partly a result of experience but, more than that, is often also an indication of extraordinary scientific insight. It should come as no surprise that Einstein provides a striking illustration of the last claim.

2. Contexts

The physical and personal backgrounds of Einstein's 1905 paper have been carefully delineated by the editors of Einstein's *Collected Papers* (1989a, pp. 206–222). Suffice it here merely to note the more salient points, without mention of which the discussion here would be incomplete: (i) Since the first report of Brownian motion in 1828, there were many attempts to explain it on physical grounds, including the use of the kinetic theory. These had all been unsuccessful. (ii) The use of the kinetic theory usually came with a commitment to atomism (only "usually," because, as in the case of Mach, it was possible to deny atomism while accepting the heuristic value of the kinetic theory). (iii) By 1900, when he finished reading Boltzmann's *Vorlesungen über Gastheorie*, Einstein had become firmly committed to atomism.[15] His only criticism of Boltzmann concerned the latter's inadequate attention to the task of connecting atomism—and the kinetic theory—to experimental data. Einstein also developed a strong interest in the foundations of the kinetic theory. (iv) Like Exner (1900) before him, who had pioneered a new method of applying the kinetic theory to the problem of Brownian motion, Einstein had come to accept the idea that there is no intrinsic difference between solutions and suspensions from the kinetic point of view. The only difference between the two involved the size of the non-solvent particle. Finally (v) Einstein remained unconvinced that the methods used for the kinetic theory of gases could be carried over—without change—into the context of liquids.

Einstein's approach to the problem of Brownian motion, unlike Exner's, does not emphasize the equipartition of energy (and, in fact, Einstein would later criticize Marian Smoluchowski's use of an equipartition assumption). Instead, Einstein starts with a relation for the osmotic pressure, bringing in as few assumptions from the underlying mechanics as possible. This leads to a recognition of a difference between classical thermodynamics and what is entailed by the kinetic theory. Einstein generally opts for the latter, insofar as he attempts to give a kinetic explanation; but, in using the osmotic pressure, he accepts the relevance, in the microscopic realm, of what is usually regarded as a concept from macroscopic thermodynamics. This step amounts to an acceptance of the idea of dissipation in a microscopic system—the model is a hybrid of classical thermodynamics and the kinetic theory. In all of these choices, Einstein showed a quite sophisticated knowledge of the physical situation. In his formal treatment, he was also inventive, but he proceeded to reinvent methods that had already been developed, though not popularized. What was unprecedented was how many of these he wove together in one argument. There are two formal antecedents particularly worthy of mention, since Einstein rediscovered methods from both.

Only one of these was in physics—a paper by Rayleigh, "Dynamical Problems in Illustration of the Theory of Gases" (1891)—in which, unlike Maxwell, Boltzmann, and others, Rayleigh considers not only the stationary states of a system, but also the truly kinetic problem of the temporal approach to them. Rayleigh analyzes a series of kinetic models designed "to bring out the fundamental statistical questions, unencumbered with other difficulties" (p. 424). Motion is restricted to one dimension, to a study of bodies "impinging" on one another, "and in addition one set of impinging bodies is supposed to be very small relatively to the other" (p. 424). For free masses, if υ is the rate of the "launch" of the smaller masses ("projectiles"), v and u the initial velocities of the smaller and larger masses, respectively, q the ratio of the smaller to the larger mass, and $f(u)du v dt$ the "number of collisions by which masses are expelled from the range du in time dt," Rayleigh derives the equation:

$$\frac{1}{\upsilon}\frac{df}{dt} = 2q^2v^2\frac{d^2f}{du^2}, \tag{1}$$

which, he observes, is the "well-known equation of the conduction of heat" (p. 433). This was probably the first demonstration that what came to be called a Wiener process (in this case, the random collisions) gives rise to a diffusion equation. The similarity to Einstein's later result (see below) is obvious, but not only did Einstein apparently not know of this paper, there is little evidence of any contemporary interest in it.

Rayleigh's work was, at least, in physics. The other striking formal antecedent of Einstein's paper was a rather remarkable thesis in economics, by Louis Bachelier (1900), which was, however, presented for a "Docteur ès Sciences mathématiques" degree, and published with a dedication to Poincaré. Bachelier's purpose was

> to study mathematically the static state of the market at a given instant, i.e., to establish the law of probability of price changes consistent with the market at that instant. If the market . . . does not predict its fluctuations, it does assess them as being more or less likely, and this likelihood can be evaluated mathematically. (Bachelier 1900, p. 17)

Let $p_{x,t}dx$ be the probability that a price is between x and $x + dx$ at time t. Bachelier wanted to calculate the "conditional probability that price z will be quoted at the moment $t_1 + t_2$, price x having been quoted at the moment t_1" (p. 29). This conditional probability, Bachelier argued, would be given by

$$\int_{-\infty}^{+\infty} p_{x_1 t_1} p_{z-x, t_2} dx dz. \tag{2}$$

This result, an almost fully general version of what came to be known as the Chapman-Kolmogorov equation, would be rediscovered (in a rather problematic form) by Einstein. Bachelier's result seems to be the first such equation to be written down.

Bachelier went on to consider the probability $\wp$ that price x is attained or exceeded at time t. He showed that this probability increases monotonically with time and that it obeyed the "Fourier equation":

$$c^2 \frac{\partial \wp}{\partial t} - \frac{\partial^2 \wp}{\partial x^2} = 0 \tag{3}$$

(where c is a constant; p. 40), thereby showing even more clearly than Rayleigh how a Wiener process gives rise to a diffusion equation. The most straightforward argument for this result, however, will also occur in Einstein's paper.

3. Einstein's Paper

The critical moves in Einstein's paper are made at the beginning of § 4. However, in order to situate that derivation in its proper context, a brief look at the earlier sections will be helpful. In § 1, Einstein points out that, whereas conventional thermodynamics made a distinction between non-electrolytes dissolved in a liquid and particles that are only suspended in it (only the former produce an osmotic pressure), "from the standpoint of the molecular-kinetic theory of heat . . . it is difficult to see why suspended bodies should not produce the same osmotic pressure as an equal number of dissolved molecules" (Einstein 1905c, p. 124). This was the move initated by Exner (1900). That osmotic pressure is calculated in § 2. In sufficiently dilute liquids, it had the same form as it did in the case of dissolved particles (assuming that all the particles are spherical).

So far, nothing is achieved beyond a rather straightforward extension of the kinetic theory of solutions to particles in suspension. § 3, however, paves the way to a more exotic destination. Einstein summarizes—without explicit reference—the results of his doctoral dissertation, *Eine neue Bestimmung der Moleküldimensionen* (Einstein 1905b), to obtain an expression for the diffusion coefficient of the suspended substance. To do so, he introduces two new approximations, as well as a hypothetical force, K, and an assumption of equilibrium between the influence of this force and the process of diffusion. Of the two approximations, one is innocuous enough, though not easily corrigible: The suspended particles are distributed at random in the liquid. The other, however, is far from innocuous, though, ironically, corrigible (see below): Referring readers to Kirchhoff (1897), Einstein invokes a formula (originally due to Stokes) for the velocity imparted by K to an individual particle:

$$\frac{K}{6\pi k P}, \tag{4}$$

where P is the radius of the (spherical) particle, and k is the coefficient of viscosity of the liquid. Now, invoking the assumption of equilibrium, Einstein equates this expression to one describing the effects of diffusion and, after some manipulation, obtains:

$$D = \frac{RT}{N}\frac{1}{6\pi kP}, \tag{5}$$

where D is the coefficient of diffusion, R the ideal gas constant, T the absolute temperature, and N Avogadro's number. This is the first important result of the paper. It sets the stage for the seminal derivation of §4.

For the purposes of this paper, the derivation in §4 must be reconstructed in some detail. Einstein begins with an assumption whose status is still problematic and certainly troubled his contemporaries: that there exists "a time interval τ, which shall be very small compared with observable time intervals but still so large that all motions performed by a particle during two consecutive time intervals τ may be considered as mutually independent events" (Einstein 1905c, p. 130).[16] This is essentially a very strong Markov postulate. Einstein makes no attempt to justify it. Instead, he proceeds to use it immediately to motivate the claim:

$$f(x,t+\tau)dx = dx \int_{\Delta=-\infty}^{\Delta=+\infty} f(x+\Delta)\,\varphi(\Delta)\,d\Delta, \tag{6}$$

where $f(x,t)$ is the number of particles per unit volume and $\varphi(\Delta)$ is defined by $dn = n\,\varphi(\Delta)\,d\Delta$ (with $\int_{-\infty}^{+\infty}\varphi(\Delta)\cdot d\Delta = 1$), where n is the number of particles, and Δ is the change in the x-coordinates of individual particles during τ (with $\varphi(\Delta) = \varphi(-\Delta)$). The notational ambiguity—f on the right-hand side has only one argument—underscores the heuristic nature of this step. As far as the second argument is concerned, there is no clear indication whether f (on the right-hand side) is to be evaluated at t or $t + \tau$. With the wisdom of hindsight, the answer is t. (Roughly, integration over Δ takes care of what happens during the subsequent τ.) The notational ambiguity hides the role of the independence (Markov) assumption, which is required to ensure that $f(t+\tau)$ (in eq. 6) does not depend on any time before t.

Since "τ is very small," Einstein then argues:

$$f(x,t+\tau) = f(x,t) + \tau\frac{\partial f}{\partial t}. \qquad (7)$$

Expanding $f(x+\Delta,t)$ in powers of Δ, he also gets:

$$f(x+\Delta,t) = f(x,t) + \Delta\frac{\partial f(x,t)}{\partial x} + \frac{\Delta^2}{2!}\frac{\partial^2 f(x,t)}{\partial x^2} \ldots \text{ ad inf.} \qquad (8)$$

Introducing yet another approximation—"We can perform this expansion under the integral since only small values of Δ contribute anything to the latter" (Einstein 1905c, p. 131)—Einstein finally obtains:

$$f + \frac{\partial f}{\partial t}\tau = f\cdot\int_{-\infty}^{+\infty}\varphi(\Delta)d\Delta + \frac{\partial f}{\partial x}\int_{-\infty}^{+\infty}\Delta\varphi(\Delta)d\Delta + \frac{\partial^2 f}{\partial x^2}\int_{-\infty}^{+\infty}\frac{\Delta^2}{2}\varphi(\Delta)d\Delta \ldots \qquad (9)$$

Now, after redefining $D = \frac{1}{\tau}\int_{-\infty}^{+\infty}\frac{\Delta^2}{2}\varphi(\Delta)d\Delta$ (and, strictly speaking, this is another new assumption) and dropping higher-order terms, he obtains "the familiar differential equation for diffusion" (Einstein 1905c, p. 132):

$$\frac{\partial f}{\partial t} = D\frac{\partial^2 f}{\partial x^2}. \qquad (10)$$

From this, a straightforward set of steps leads to his best-known result:

$$\lambda_x = \sqrt{\overline{x^2}} = \sqrt{2Dt} = \sqrt{t}\sqrt{\frac{RT}{N}\frac{1}{3\pi kP}}. \qquad (11)$$

If N is assumed to be known, λ_x can be calculated; alternatively, if λ_x is measured, then this relation provides a recipe for determining N.

4. The Reception

The story of the great interest that Einstein's paper aroused in the physics community is well-known (see Einstein 1989a, pp. 206–222). Suffice it here—as in the discussion of the physical origins of this work—to note only those important developments that are particularly relevant to the themes of this paper:

(i) By 1910, a series of brilliant experiments by Perrin and his students had provided ample experimental confirmation of Einstein's results. By the time this work was completed there was little room left to doubt the validity of atomism (see Nye 1972). By about 1907, Ostwald had capitulated; by 1910 even Mach seems to have become more amenable to atomism (see Einstein 1989a, p. 218, and Blackmore 1972).

(ii) Within physics, two important papers were written in an attempt to provide an alternative foundation for Einstein's result (eq. 11).[17] The more important of these, the paper by Langevin (1908), will be considered in some detail below. A less important one—though one often hailed as a significant advance over Einstein's derivation—was published by Smoluchowski (1906b), who had started work on Brownian motion independently of Einstein. Smoluchowski's later reputation for systematically developing the theory of Brownian motion is largely due to his later work, done after 1910.[18] The 1906 paper is less than flattering to Smoluchowski's reputation. Smoluchowski claims to be offering a method "more direct, simpler, and thus perhaps also more convincing than that of Einstein" (Smoluchowski 1906b, p. 756; as quoted in Einstein 1989a, p. 216). However, the paper eschews systematic derivations altogether—even at Einstein's level of rigor—and replaces them by relatively mundane heuristic arguments, arriving at a result for $\overline{x^2}$ that differed from Einstein's result by a factor of 64/27. Langevin (see below) would attempt to remove this discrepancy. Moreover, as Einstein later points out (in 1917), Smoluchowski arrives at his result by applying the equipartition theorem to the velocity of a suspended particle, and by making the special assumption that "this velocity is constantly destroyed by internal friction and constantly recreated by irregular molecular impulses" (Einstein 1917, p. 737; as quoted in Einstein 1989a, p. 216). This assumption, Einstein correctly points out, is far more restrictive than the assumption about dissipation (that is, about the existence of an osmotic pressure) that he, himself, had used. Even though Smoluchowski had attempted to describe Brownian motion in both gases and liquids—an ostensibly more general situation than that considered by Einstein—the latter also felt that Smoluchowski's work only applied to gases (see Einstein 1989a, p. 217).

(iii) Finally, Einstein's paper marks the beginning of the thermodynamic treatment of fluctuation phenomena, which Einstein interpreted as a departure from classical thermodynamics (and, therefore, a challenge to the energeticists). Einstein pursued this development further in 1910 while computing the scattering of light in the atmosphere (Einstein 1910). This paper formed the basis for later developments in the thermodynamics of fluctuations (see Tisza 1966, pp. 296–306).

Two approximations that Einstein used received immediate systematic attention. The first was Stokes's formula (eq. 4). Stokes's derivation of that result had assumed that liquid particles adhere completely to the surface of the suspended one, so that the latter's velocity is negligible, while the hydrodynamic (Navier-Stokes) equations continue to hold. At best, Stokes's formula can only be regarded as (corrigibly) approximate, and the conditions of its derivation are far from satisfied in the context where Einstein invoked it. However, a series of investigations attempted to provide corrections (see Fürth 1922, pp. 52–56) whose effects could be at least implicitly estimated. For the range of velocities relevant to Brownian motion, it emerged unscathed. Though apparently so ad hoc, and the result of a derivation based almost entirely on counterfactual assumptions, Stokes's formula provides a textbook case of unproblematic approximation.

Einstein's introduction of τ, however, had no such resolution. Attempts to remove this approximation involved entirely new derivations of eq. 11—the principal focus of interest. These can only be regarded as "corrections" in a rather Pickwickian sense. Following work by Leonard Ornstein and Johannes Wander de Haas-Lorentz (see Fürth 1922, pp. 97–98), Fürth derived the formula:

$$\overline{x^2} = 2D(t - mB + e^{-t/mB}), \tag{12}$$

where $B = 1/(6\pi kP)$ and m is the mass of the particle, with the other symbols being the same as before (Fürth 1920). This equation reduces to eq. 11 only when $t \to \infty$. For $t \sim 0$, the mean fluctuation is independent of time. This precludes Einstein's derivation from emerging as a limit of *any* derivation of eq. 12.

However, in Einstein's derivation, there is no easy way out of this approximation. What it amounts to is an assumption that the motion continues to be described by a Markov process even when modeled with infinitesimally small time-intervals—a counterfactual physical approximation. More formally, Einstein's model assumes a discrete series of instants (of time) $0, \tau, 2\tau, 3\tau, \ldots$ at which impacts take place, as well as assuming that the process is Markovian and that the impulses felt are random; eq. 6 defines this model. (From this point of view, the diffusion equation [eq. 10] is an approximate description of this process and its derivation is to be regarded as nothing more than a motivation for the approximation.) What Fürth's—or any other such—derivation amounts to is the formulation of a different model for Brownian motion, and (in Fürth's case) one that is only barely compatible with Einstein's. Langevin (see below) would provide a much more interesting—at least much more fecund—model that is more compatible with it.

Eq. 6 is, of course, a version of the Chapman-Kolmogorov equation that Bachelier had already written down. It forms the critical equation on which the theory of Markov processes is based, and which began to receive systematic attention after the publication of the German edition of Markov's *Wahrscheinlichkeitsrechnung* (Markov 1912). In 1913, Fokker began the systematic development of diffusion equations (of the form of eq. 10); in this case, Einstein's inspiration was direct (see Fokker 1914).[19] Further mathematical development of the type of expansion that Einstein used to get from the Chapman-Kolmogorov to the diffusion equation (from eq. 6 to eq. 10), which came to be known as the Kramers-Moyal expansion (see Moyal 1949), came later. In effect, this part of Einstein's derivation incorporated the major themes that would dominate the study of Markov models during the next fifty years.[20] What Einstein had done, apparently with only mathematical and physical intuition as a guide, was to carry out a sequence of innovative steps, each of which had some antecedents, but which, as a whole, were startlingly modern—and this is only a remark about the formal derivation, leaving aside everything that Einstein's treatment of Brownian motion did for physics.

5. Random Walks

In the development of a mathematical theory of stochastic processes one other important formal development was taking place almost simultaneously with Einstein's work on Brownian motion. While working on a purely biometrical problem, Pearson introduced a formal "problem of the random walk" to the readers of *Nature*:

> A man starts from a point O and walks l yards in a straight line; he then turns through any angle whatever and walks l yards in a second straight line. He repeats this process n times. I require the probability that after these n stretches, he is at a distance between ρ and $\rho + \delta\rho$ from his starting point, O. . . . A solution ought to be found, if only in the form of a series in powers of $1/n$, when n is large. (Pearson 1905a, p. 294)

In the very next issue of *Nature*, Rayleigh (1905) pointed out that Pearson's problem was "the same as that of the composition of n isoperiodic vibrations of unit amplitude and phases distributed at random," for which he had provided an asymptotic (large n) solution in closed form as early as 1880:

$$\frac{2}{n} e^{-r^2/n} r dr \tag{13}$$

(Rayleigh 1880). Pearson (1905b) acknowledged this help, and that of G.J. Bennett, who had solved the case of $n = 3$.[21] In 1894, however, Rayleigh had gone further. Exploiting the technique for converting a problem of determining a probability distribution function that he had used earlier (Rayleigh 1891; see above), Rayleigh had converted the formal problem into that of solving a diffusion equation (Rayleigh 1894, §42a) and, in 1899, he extended the method (Rayleigh 1899).

In 1905, J.C. Kluyver pursued this strategy of analysis systematically (Kluyver 1905), as did Pearson in 1906, with his characteristic attention to detail (Pearson 1906). Rayleigh returned to the problem in 1919, and provided a comprehensive—and comparative—analysis of the random walk problem in one, two, and three dimensions (Rayleigh 1919). Rayleigh's treatment was eventually further generalized by Subrahmanyan Chandrasekhar (1943). Somewhat surprisingly, none of them made an explicit connection between Einstein's model of Brownian motion and the random walk problem, even though the differential equation obtained was the same. Kluyver and Pearson, perhaps understandably, seem to have been unaware of that development. Rayleigh seems not to have noticed the connection, and Chadrasekhar, though certainly aware of the connection, used Langevin's model (see below), rather than Einstein's, as the starting point of his analysis of Brownian motion because of its greater potential for generalization.[22]

Independent of Rayleigh's early work, and that of Kluyver and Pearson, Smoluchowski (1906a) also provided a complete solution of the one-dimensional case, though without a transition to the differential equation. As noted before, Smoluchowski's (1906b) paper on Brownian motion also did not involve that transition. Though the problem of random walks provides one of the most popular points of entry for Brownian motion into the theory of stochastic processes (see, for example, Kac 1947), explicit recognition of that connection came slowly. However, the earliest clear (though only implicit) statement of that connection is in a paper by Einstein himself.

In 1908, at the request of Richard Lorenz, who "pointed out . . . that many chemists would welcome an elementary theory of Brownian motion" (Einstein 1989a, p. 218), Einstein published a semi-popular exposition of his theory, with heuristic arguments (though of no less rigor than Smoluchowski 1906b) replacing the derivation of his 1905 paper (Einstein 1908). For the motion of solute molecules (or suspended particles—for Einstein, there is no essential difference), Einstein offered an analysis that is clearly that of a random walk in one dimension:

> We imagine that we know the x-coordinates of all dissolved molecules at a certain time t, and also at time $t + \tau$, where τ denotes a time interval so short that the concentrations in our solution change very little during it. During this time τ, the x-coordinate of the first dissolved molecule will change by a certain quantity Δ_1 on account of the random thermal motion, that of the second molecule will change by Δ_2, etc. These displacements, Δ_1, Δ_2, etc., will be in part negative (directed to the left) and in part positive (directed to the right). Furthermore the magnitude of these displacements will vary from molecule to molecule. But since we assume . . . that the solution is dilute, this displacement is determined only by the surrounding solvent, while the rest of the dissolved molecules have no appreciable effect; for that reason these displacements Δ will *on the average* be of equal magnitude in parts of the solution having differing concentrations, and will be just as often positive as negative. (Einstein 1908, p. 237)

Through the medium of Perrin (1910),[23] this approach filtered down to Wiener (1923), who describes the problem of Brownian motion in almost identical terms in the first of a series of papers that established the mathematical foundations for the type of derivation that Bachelier, Einstein and, later, Fokker and many others had given.

6. Langevin's Model

If Smoluchowski's derivation already introduces important physical differences from Einstein's, and thereby presents a different model of Brownian motion, a third derivation, due to Langevin, introduces not only a new physical model, but (in that process) a radically new mathematical *Ansatz* of immense fecundity. Langevin claimed to be providing a "correct application" of Smoluchowski's method, thereby removing the numerical discrepancy between his and Einstein's results, and also providing a proof that was "infinitely more simple" ["une démonstration infiniment plus simple"] (Langevin 1908, p. 530) than that of Einstein.

But for its "novel" mathematics, the proof is indeed simple. If ξ is the velocity of a Brownian particle, m its mass, R the ideal gas constant, T the absolute temperature, and N Avogadro's number, then, at equilibrium:

$$m\xi^2 = \frac{RT}{N}. \tag{14}$$

Langevin assumes that the particle is subject to two sorts of forces: (i) a viscous drag given by $-6\pi\mu a\xi$, where μ is the viscosity and a the diameter of the particle; and (ii) an "irregular" force, X, about which nothing is known except that (in any given direction, say x,) it is as likely to be positive as negative (Langevin 1908, p. 531). These assumptions give rise to what is known as Langevin's equation:

$$m\frac{d^2x}{dt^2} = -6\pi\mu a\frac{dx}{dt} + X. \tag{15}$$

Multiplying both sides by x,

$$\frac{m}{2}\frac{d^2x^2}{dt^2} - m\xi^2 = -3\pi\mu a\frac{dx^2}{dt} + Xx. \tag{16}$$

Now, taking an average over a large number of particles, using eq. 14, and setting $\overline{Xx} = 0$ because of the "irregularity" of X (Langevin 1908, p. 532):

$$\frac{m}{2}\frac{dz}{dt} + 3\pi\mu a z = \frac{RT}{N}, \tag{17}$$

with $z = \dfrac{\overline{dx^2}}{dt}$. The general solution of eq. 17 is given by:

$$z = \frac{RT}{N}\frac{1}{3\pi\mu a} + Ce^{-6(\pi\mu a/m)t} \tag{18}$$

where C is a constant. Langevin estimates that the second (decaying exponential) term becomes negligible in 10^{-8} seconds and could, therefore, be neglected, leaving:

$$\frac{\overline{dx^2}}{dt} = \frac{RT}{N}\frac{1}{3\pi\mu a}. \tag{19}$$

This is now solved to give Einstein's relation (eq. 11). Langevin's analysis has no place for Einstein's τ. It is a beautiful derivation, provided that one is willing to countenance the introduction of X, and its subsequent removal. It remains somewhat mysterious why Langevin viewed it as an application of Smoluchowski's method, which was far less formally ingenious besides making some doubtful physical assumptions.

Note that the move from eq. 18 to eq. 19 involved a standard approximation—dropping a term on physical grounds, and that this is necessary to get agreement with Einstein's result. However, this is a corrigible approximation and, arguably, does not introduce any new assumption. Langevin's crucial innovation is eq. 15. This is not a differential equation in any ordinary sense: Because of the unusual (fluctuating) properties of X, it cannot be straightforwardly integrated. Historically, eq. 15 was the first "stochastic differential equation." While Langevin's equation became a standard point of departure for the development of models of "Brownian motion" subject to various conditions (such as those of a particle subject to various fields of force, see Chandrasekhar 1943), the equation had to come with a warning about its special status.

Uhlenbeck and Ornstein (1930), for instance, had to introduce special assumptions about the integral of *X*; these are mathematical steps (or, rather, restrictions) based on physical intuitions. In his famous review of stochastic models in physics and astrophysics, Chandrasekhar was even more explicit:

> . . . "solving" a stochastic differential equation like [Langevin's equation] is not the same thing as solving any ordinary differential equation. [X] has only statistically defined properties. Consequently, "solving" the Langevin [equation] has to be understood rather in the sense of specifying a probability distribution. (Chandrasekhar 1943, p. 21)

The mathematical framework for the analysis of these equations was cleared up by Itô only in the 1940s (see, especially, Itô 1951); the result is a radically new (and beautiful) mathematical theory, the Itô calculus, which had hardly even been intuited by Langevin and those who had followed him during the next three decades.

We return now to the question of what physical approximations mean. Einstein and Smoluchowski had differed on points of physics, both explicit and implicit. Since there is an explicit discrepancy between their results, however small that may be, one could conclude that they simply had, at least weakly, different (and incompatible) models, and that only one of these models could be correct. The relation between Einstein's and Langevin's analyses, however, is more problematic—and interesting. Formally, Einstein had introduced "new physics" (in the sense that the steps were not all "physically justified," given what had been explicitly assumed), whereas Langevin had done more than that. He had also introduced what amounts to "new mathematics" on the basis of physical plausibility. (In that sense, Langevin's move was not "mathematically justified.") The last few remarks should, however, be treated with some caution. While it is clear that both Einstein and Langevin introduce new (and different) assumptions of a mathematical form, the claim that Einstein had introduced what is primarily new physics (in eq. 6) is at least not transparent, for where mathematics ends and physics begins is far from clear in these models.

Leave aside the question what, exactly, is physics and what is only mathematics in these models, which is probably of interest only to professional philosophers schooled in a firm distinction between mathematics and physics (and the concomitant doctrines of the absence of empirical content in mathematics).[24] The critical point is that, whereas Einstein and Langevin agree on eq. 11, there is no translation procedure, so to say, that would transform one derivation into the other. This requires some explanation and that, in turn, is not hard; one has to tell a story of the following sort (see Gardiner 1985, p. 7). Langevin's averaging procedure, which leads to $\overline{Xx}$ being equal to 0, actually requires two separate assumptions: (i) the irregularity of X; and (ii) the independence (in a statistical sense) of X and x. These assumptions are, in turn, equivalent to Einstein's independence postulate (the one that led eq. 6 to be written down) in the sense that they play the same role in Langevin's derivation. The irregularities of X (as a function of time) must be strong enough that they do not somehow conspire to give $\overline{Xx}$ a non-zero value.

The "philosophical" trouble with this explanation is that, in spite of the obvious insight that it offers, it still does not permit the sort of transformation just mentioned; the explanation remains verbal, and, at present, no more formal explanation is forthcoming.[25] What the explanation does—and it does no more—is to establish, in some sense, the *compatibility* of the models (one that was already suggested by the agreement with respect to eq. 11). If Einstein's model is taken as the point of departure, Langevin's model presents approximations that are implicit and incorrigible (probably in principle), lack any known procedure for estimating all their effects, involve an (implicit) step that is neither physically nor mathematically justified (from the standpoint of the physics and mathematics assumed prior to that step), and are probably context-dependent and certainly counterfactual. Conversely, if Langevin's model is taken as the point of departure, Einstein's model would "suffer" from most of these "problems"! There should be little doubt, though, that both represent important advances in physics and mathematics.

This begs a question: "Which of these is the *real* Brownian motion?" If the kind of realism that would be called "external realism" is to be adopted (to use an old distinction of Carnap's; see Carnap 1950), there would be an answer: The *real* Brownian motion is the one independent of all these models. The flaws with such a position are too well known to be elaborated here. The other option is "internal realism." From this point of view the ontology is that of both entities and processes within a particular model ("formal system" in philosophers' jargon). The interesting point is that, if internal realism is adopted, then, since both models give rise to the same equation at the end (eq. 11), since neither violates any very general physical principles, and since both are compatible with each other, but are nevertheless not transformable into one another, one is left with divergent models and, therefore, diverging ontologies (of entities and processes) even in the case of a well-worn example of physics.[26] Once the many other models of "Brownian motion" that populate stochastic process theory are also brought into the picture, this diversity becomes further enhanced. Recourse to observation ("what Brown saw") is of no help. Beyond eq. 11, all that would remain are rather vague reports of irregular motions of suspended particles in fluids. In any case, agreement about these is hardly agreement about the ontology of theoretical physics.[27] But all of this is probably no "real" reason to worry as long as one is willing to forego the desert of doctrinaire physicalism (with a unifying, single, consistent physical language, that of theoretical physics) for the lush ontology of a tropical forest.[28] One could also avoid these moves altogether by proceeding to do science as scientists do, and refusing to be involved in philosophical disputes arising from conflicting manifestoes of ontological intolerance.

Acknowledgements. It is a pleasure to acknowledge many illuminating conversations with John Stachel, the influence of Abner Shimony (especially his constant reminder that the philosophy of physics should not remain constrained to the axiomatics of quantum mechanics and space-time theories) and, finally, the encouragement of William Wimsatt, who has long championed the view that an understanding of the process of approximation is critical for the philosophy of science. Don Howard commented extensively on an earlier draft of this paper—his comments led to many clarifications and changes. The work on which this paper is based was supported by a Fellowship from the Dibner Institute at MIT.

NOTES

[1] That paper is dated May 1905 (received on 11 May 1905 at the *Annalen der Physik*), in between the other two great papers, those on light quanta, Einstein 1905a (dated 17 March 1905), and special relativity, Einstein 1905d (dated June 1905); for publication details, see Einstein 1989a.

[2] Leggett makes these remarks in the context of condensed matter physics and does not indicate their relevance elsewhere. Their use in the text in a much more general context, therefore, goes beyond what Leggett explicitly says; however, there is no reason why they should not be taken to have this more general import. In any case, the particular example that is being discussed here is close enough to Leggett's concerns in the sense that it also concerns a transition from the microscopic to the macroscopic. An effort will be made throughout this paper to restrict examples used to such cases. (Because of this feature, this paper is also about what philosophers call "reductionism." However, the issues connected with reductionism will not be explicitly analyzed here; on that point, see Sarkar 1992.)

[3] See, however, Cartwright 1983 for views somewhat more akin to those of Leggett (and the ones being advocated in this paper).

[4] In a biological context, a somewhat similar point of view was developed, very early, by Levins (1966) and elaborated by Wimsatt (see, for example, Wimsatt 1986, 1987). Both of them, however, implicitly endorse some form of (weak) realism, and concentrate on models with counterfactual approximations (see below). Assuming realism has the consequence that adding details to various models should ensure their eventual convergence. Leggett takes no explicit position on that issue though the tenor of his remarks suggests some skepticism. This kind of realism is *not* being endorsed here and, as will be obvious from the discussion in the text, no assumption is made about the convergence of approximations. It should also be emphasized that the similarity between the views being discussed here and the so-called "semantic interpretations of theories" is superficial, and does not go beyond the recognition of the rather uncontroversial fact that models are the daily currency of scientific research.

[5] The distinction, here, is between heuristics used directly to write down a formula (as, for instance, in Smoluchowski 1906b), which is rarely fecund, and heuristics used to transform formulae, as in derivations. It is the latter that makes Einstein's paper so important.

[6] Note that there may be some room for disagreement, even on the last point. This paper follows the usual convention in saying that Einstein's paper is about "Brownian motion." Einstein, however, was more guarded in 1905: "It is possible that the motions to be discussed here are identical with the so-called 'Brownian molecular motion'; however, the data available to me on the latter are so imprecise that I could not form a definite opinion on this matter" (Einstein 1905c, p. 123). There is, however, no reason to believe that this caution reflects any genuine doubt on Einstein's part.

[7] This paper, however, is not an attempt to use Einstein's example as evidence for the philosophical thesis about the ubiquity of physical approximations that was mentioned above. Ahistorical, reconstructed science, often (at least) implicitly reconstructed with a philosophical thesis already in mind, is usually dubious as evidence for that thesis. Rather, what this paper attempts to do is to use the philosophical analysis that will be developed to illuminate a historical example, Einstein's early work on Brownian motion and its relation to its intellectual context. Of course, should such an illumination offer new insight, that would constitute indirect evidence in favor of that thesis, but it will not be interpreted here in such a fashion. The point of view that is endorsed here is that philosophical theses about scientific method must find their justification in contemporary science, that is, in their ability to provide benefits for actual scientific practice. Historical examples may provide sources of insight for epistemological theses of this sort, but reconstructed science is questionable as evidence for them.

[8] Surprisingly, this story is almost completely ignored in the otherwise detailed and illuminating editorial comments in Einstein 1989a. A history of the development of stochastic processes remains to be written and the partial sketch

given here is no more than an indication of historical problems that deserve further attention.

[9] Obviously, there is something to be said for keeping approximations explicit: that makes it easier to gauge the effects of the approximation. But a stricture that all approximations should always be explicit would prove cumbersome in many physical contexts. The socialization of physicists guards against errors from most common implicit assumptions.

[10] For instance, in most models in the (classical) kinetic theory of gases, an incorrigible approximation—the spherical shape of the gas molecules—is only incorrigible in practice. Einstein will routinely use this approximation. In models of phase transitions (or many other models of quantum statistical mechanics), or models involving sensitive dependence on initial conditions, the incorrigibility is one of principle.

[11] It should be emphasized that "new physics" does not mean some new, putatively universal principles. Rather, what matters here are assumptions at some level that can be only slightly removed from the so-called phenomenological laws. As long as these assumptions are not clearly derivable from the physics that has already been explicitly assumed, they are "new" in the relevant sense. As Shimony (1987) has observed, in many derivations that ostensibly reduce some physical phenomenon to "deeper" laws, the approximations that are used involve (usually subtly) assumptions at the level of what is apparently being reduced. The putative reductions then become problematic: in the terminology of this paper, new physics has been brought in.

[12] The trouble is that classical thermodynamics requires the linear increase of all extensive variables, including the energy, with size. This precludes the constituents of the systems from interacting with any long-range interaction (gravitation, electrostatics, *etc*). In principle, the resulting approximations should be corrigible —there is nothing puzzling about the physics—though no one has so far provided a careful analysis of this situation.

[13] The ubiquity of counterfactual assumptions raises obvious questions about ontology (at least, about physicalism). One response would be to assume that all theories are approximations, and that the "underlying world" poses no problem—in effect, use instrumentalism to rescue realism. Another response, which is Leggett's implicit move, is to admit that these counterfactual assumptions should be regarded as new physical hypotheses of the special sort indicated in the text.

[14] On this point, see also Shimony 1987, a paper that is one of a very few philosophical attempts to address the actual complexity of the "ontology" of physics.

[15] Don Howard (personal communication) has pointed out that, though Einstein later described the establishment of atomism as a goal of this early work, the work does not make any significant assumption about the exact properties of atoms. All that is required of atomism, at this stage, is that the system have a finite number of degrees of freedom.

[16] As Fürth put it, in 1922, in his edition of Einstein's papers on Brownian motion: "The introduction of this time-interval τ forms a weak point in Einstein's argument, since it is not previously established that such a time-interval can be assumed at all. For it might well be the case that, in the observed interval of time, there was a definite dependence of the motion of the particle on the initial state" (Fürth 1922, p. 97).

[17] Attention was generally restricted to eq. 11 because of its experimental import. On the one hand, this ensured that the new work—if it resulted in eq. 11—would be compatible with Einstein's results (in this sense). On the other hand, eq. 11 is the result of a rather unique derivation. The new derivations largely ignored this point and, with the exception of Langevin 1908 and Smoluchowski's *later* work, were still-born in the sense that they led to no further development.

[18] See Sommerfeld 1917 for a comprehensive account.

[19] This family of equations came to be known variously as the Fokker-Planck equations, the forward diffusion equations, or the forward Kolmogorov equations. Note that Fokker's direct inspiration was Einstein's work on radiation, rather than Brownian motion; the former had led to more complicated diffusion equations.

[20] See, also, Gardiner 1985, pp. 2–6.

[21] Meanwhile yet another *Nature* correspondent, Fessenden (1907), attempted to illuminate the problem by pointing out its use by Kipling in "The Strange Ride of Morrowbie Jukes." The hero's directions for navigating a quicksand were "Four out from crow-clump; three left; nine out; two right; three back; two left; fourteen out; two left; seven out; one left; nine back; two right; six back; four right; seven back"—presumably a random walk!

[22] As Chandrasekhar puts it: "In Einstein's and Smoluchowski's treatment of the problem, Brownian motion is idealized as a problem of random flights; but . . . this idealization is valid only when we ignore effects which occur in time intervals of order β^{-1}. For the general treatment of the problem we require our discussion to be based on an equation of the type first introduced by Langevin" (Chandrasekhar 1943, p. 87, n. 18). Note, however, that Einstein had not explicitly mentioned the problem of the random walk ("random flights").

[23] Wiener does not seem to have known Einstein's papers at first hand. He only refers to Perrin as his source for Einstein's results.

[24] Turning, briefly, and only in the footnotes, to these justly-obscure philosophical disputes, note that this distinction is not identical with the "analytic-synthetic" distinction criticized by Quine (1950). The putative mathematics-physics distinction is a non-linguistic distinction and, leaving aside the doubtful cogency of Quine's strictures against analyticity (see Stein 1992 on this point), this distinction may be palatable to many who would otherwise follow Quine (on the question of analyticity). For instance, a sort of logicism that saw mathematics in continuity with logic would enable an adherent to use a distinction between logical truth and non-logical fact (which Quine sometimes admits) to argue for a distinction between mathematics and physics. (Quine, of course, finds both distinctions distasteful.) Nevertheless, what examples such as the one being discussed show is that, in

practice, the distinction between mathematics and physics is far from clear and, more importantly, devoid of scientific interest. This point was anticipated in part, long ago, by Mac Lane (1938), in his review of Carnap's *Logical Syntax of Language*.

[25] This does not detract at all from the value of the explanation—after all, it would be more than odd if explanations that obviously help understanding were devalued because of a failure to fit some pre-ordained philosophical formalism. In fact, it should suggest that the philosophers' sense of "explanation" might well be wanting. To press this point further, satisfaction of the formal criteria should, by itself, not be indicative of the success of an explanation unless there is at least some sense in which that explanation, at the level of "understanding," provides a better picture of the events.

[26] Carnap's escape, of course, is not to compare the two models beyond their experimental consequences. For those prone to talk about ontology (and, to be just, Carnap was not one of them), this kind of disciplined asceticism is almost impossible. The solution, as Carnap would probably have agreed, is to de-emphasize ontological issues and to emphasize epistemological ones. Only if one is forced—by weakness, habit, indiscretion, or academic environment—to treat ontology as important, the rest of this paragraph (in the text) becomes relevant.

[27] Even doctrinaire anti-realists (such as van Fraassen and his followers) could happily live with the incontrovertible motion of these suspended particles.

[28] The "tropical forest" metaphor is due to Wimsatt. The implausibility of physicalism—Carnap's or Quine's—is argued in detail in Sarkar (1995).

REFERENCES

Bachelier, Louis (1900). *Théorie de la Spéculation*. Paris: Gauthier-Villars. Reprinted in *The Random Character of Stock Market Prices*. Paul H. Cootner, ed. Cambridge, Massachusetts: MIT Press, 1964, pp. 17–78.

Blackmore, John T. (1972). *Ernst Mach: His Work, Life, and Influence*. Berkeley: University of California Press.

Carnap, Rudolf (1950). "Empiricism, Semantics, and Ontology." *Revue Internationale de Philosophie* 11: 20–40.

Cartwright, Nancy (1983). *How the Laws of Physics Lie*. Oxford: Oxford University Press.

Chandrasekhar, Subrahmanyan (1943). "Stochastic Problems in Physics and Astronomy." *Reviews of Modern Physics* 15: 1–89.

Einstein, Albert (1905a). "Über einen die Erzeugung und Verwandlung des Lichtes betreffenden heuristischen Gesichtspunkt." *Annalen der Physik* 17: 132–148.

— (1905b). *Eine neue Bestimmung der Moleküldimensionen*. Bern: K.J. Wyss. Revised version published as "Eine neue Bestimmung der Moleküldimensionen." *Annalen der Physik* 19 (1906): 289–305.

— (1905c). "Über die von der molekularkinetischen Theorie der Wärme geforderte Bewegung von in ruhenden Flüssigkeiten suspendierten Teilchen." *Annalen der Physik* 17: 549–560. Page numbers and quotations taken from the English translation in Einstein 1989b, pp. 123–134.

— (1905d). "Zur Elektrodynamik bewegter Körper." *Annalen der Physik* 17: 891–921.

— (1908). "Elementare Theorie der Brownschen Bewegung." *Zeitschrift für Elektrochemie und angewandte physikalische Chemie* 14: 235–239. Quotations are taken from the English translation in Einstein 1989b, pp. 318–328.

— (1910). "Theorie der Opaleszenz von homogenen Flüssigkeiten und Flüssigkeiten in der Nähe des kritischen Zustandes." *Annalen der Physik* 33: 1275–1298.

— (1917). "Marian von Smoluchowski." *Die Naturwissenschaften* 5: 737–738.

— (1922). *Untersuchungen über die Theorie der 'Brownschen Bewegung'.* Reinhold Fürth, ed. Leipzig: Akademische Verlagsgesellschaft.

— (1926). *Investigations on the Theory of the Brownian Movement.* Reinhold Fürth, ed. A.D. Cowper, trans. London: Methuen; reprinted New York: Dover, 1956.

— (1989a). *The Collected Papers of Albert Einstein.* Vol. 2, *The Swiss Years: Writings, 1900–1909.* John Stachel, et al., eds. Princeton: Princeton University Press.

— (1989b). *The Collected Papers of Albert Einstein.* Vol. 2, *The Swiss Years: Writings, 1900–1909. English Translation.* Anna Beck, trans. Peter Havas, consultant. Princeton: Princeton University Press.

Exner, Franz (1900). "Notiz zu Brown's Molecularbewegung." *Annalen der Physik* 2: 843–847.

Fessenden, Reginald Aubrey (1907). "The Problem of the Random Path." *Nature* 75: 392.

Fokker, Adriann Daniel (1914). "Die mittlere Energie rotierender elektrischer Dipole im Strahlungsfeld." *Annalen der Physik* 43: 810–820.

Fürth, Reinhold (1920). "Die Brownsche Bewegung bei Berücksichtigung einer Persistenz der Bewegungsrichtung. Mit Anwendungen auf die Bewegung lebender Infusioren." *Zeitschrift für Physik* 2: 244–256.

— (1922) "Anmerkungen." In Einstein 1922, pp. 54–72. Page numbers and quotations taken from the English translation as "Notes" in Einstein 1926, pp. 86–119.

Gardiner, Crispin W. (1985). *Handbook of Stochastic Methods for Physics, Chemistry, and the Natural Sciences*, 2nd. ed. Berlin: Springer-Verlag.

Itô, Kiyoshi. (1951). "On Stochastic Differential Equations." *Memoirs of the American Mathematical Society* 4: 1–51.

Kac, Mark (1947). "Random Walk and the Theory of Brownian Motion." *American Mathematical Monthly* 54: 369–391.

Kirchoff, Gustav Robert (1897). *Vorlesungen über mathematische Physik*. Vol. 1, *Mechanik*, 4th ed. Leipzig: B.G. Teubner.
Kluyver, J.C. (1905) "Een vraagstuk van meetkundige waarschijnlijkheid." *Koninklijke Akademie van Wetenschappen te Amsterdam. Wisen Natuurkundige Afdeeling. Verslagen van de Gewone Vergaderingen* 14 (1905–06): 325–334. (Meeting of 30 September 1905). Reprinted in translation as: "A local probability problem." *Koninklijke Akademie van Wetenschappen te Amsterdam. Section of Sciences. Proceedings* 8 (1905): 341–350.
Langevin, Paul (1908). "Sur la Théorie du mouvement Brownien." *Académie des Sciences* (Paris). *Comptes Rendus* 146: 530–533.
Leggett, A.J. (1987). *The Problems of Physics*. Oxford: Oxford University Press.
Levins, Richard (1966). "The Strategy of Model Building in Population Biology." *American Scienctist* 54: 421–431.
Mac Lane, Saunders (1938). "Carnap on Logical Syntax." *Bulletin of the American Mathematical Society* 44: 171–176.
Markov, Andrei Andreevich (1912). *Wahrscheinlichkeitsrechnung*. Heinrich Liebmann, trans. Leipzig and Berlin: B.G. Teubner.
Moyal, J.E. (1949). "Stochastic Processes and Statistical Physics." *Journal of the Royal Statistical Society* B11: 150–210.
Nye, Mary Jo (1972). *Molecular Reality: A Perspective on the Scientific Work of Jean Perrin*. New York: American Elsevier.
Pearson, Karl (1905a). "The Problem of the Random Walk." *Nature* 72: 294.
— (1905b). "The Problem of the Random Walk." *Nature* 72: 342.
— (1906). *Drapers' Company Research Memoirs*. Biometric Series III. London: Department of Applied Mathematics, University College, London.
Perrin, Jean (1910). *Brownian Movement and Molecular Reality*. Frederick Soddy, trans. London: Taylor and Francis.
Quine, Willard Van Orman (1950). "Two Dogmas of Empiricism." *Philosophical Review* 60: 20–43.
Rayleigh, Lord [Strutt, John William] (1880). "On the Resultant of a Large Number of Vibrations of the Same Pitch and of Arbitrary Phase." *Philosophical Magazine* 10: 73–78.
— (1891). "Dynamical Problems in Illustration of the Theory of Gases." *Philosophical Magazine* 32: 424–445.
— (1894). *The Theory of Sound*, 2nd. ed. London: MacMillan.
— (1899). "On James Bernoulli's Theorem in Probabilities." *Philosophical Magazine* 47: 246–251.
— (1905). "The Problem of the Random Walk." *Nature* 72: 318.
— (1919). "On the Problem of Random Vibrations, and of Random Flights in One, Two, or Three Dimensions." *Philosophical Magazine* 37: 321–347.
Sarkar, Sahotra (1992). "Models of Reduction and Categories of Reductionism." *Synthese* 91: 167–194.
— (1995). "Beyond Physicalism." Preprint, Department of Philosophy, McGill University.

Shimony, Abner (1987). "The Methodology of Synthesis: Parts and Wholes in Low-Energy Physics." In *Kelvin's Baltimore Lectures and Modern Theoretical Physics*. Robert Kargon and Peter Achinstein, eds. Cambridge, Massachusetts: MIT Press, pp. 399–423.

Smoluchowski, Marian (1906a). "Sur le chemin moyen parcouru par les molécules d'un gaz et sur son rapport avec la théorie de la diffusion." *Academie des Sciences de Cracovie. Bulletin International*: 202–213.

— (1906b). "Zur kinetischen Theorie der Brownschen Molekularbewegung und der Suspensionen." *Annalen der Physik* 21: 626–641.

Sommerfeld, Arnold (1917). "Zum Andenken an Marian von Smoluchowski." *Physikalische Zeitschrift* 18: 533–539.

Stein, Howard (1992). "Was Carnap Entirely Wrong, After All?" *Synthese* 93: 275–295.

Thompson, Colin J. (1972). *Mathematical Statistical Mechanics*. New York: MacMillan.

Tisza, Laszlo (1966). *Generalized Thermodynamics*. Cambridge, Massachusetts: MIT Press.

Uhlenbeck, George E. and Ornstein, Leonard S. (1930). "On the Theory of the Brownian Motion." *Physical Review* 36: 823–841.

Wiener, Norbert (1923). "Differential Space." *Journal of Mathematics and Physics* 2: 131–174.

Wimsatt, William C. (1986). "Heuristics and the Study of Human Behavior." In *Metatheory in Social Science: Pluralisms and Subjectivities*. Donald W. Fiske and Richard A. Shweder, eds. Chicago: University of Chicago Press, pp. 293–314.

— (1987). "False Models as Means to Truer Theories." In *Neutral Models in Biology*. Matthew H. Nitecki and Antoni Hoffman, eds. New York: Oxford University Press, pp. 23–55.

Einstein's Light-Quantum Hypothesis, or Why Didn't Einstein Propose a Quantum Gas a Decade-and-a-Half Earlier?

John Stachel

1. Introduction

In his second paper "On the Quantum Theory of the Monatomic Ideal Gas," Einstein proposes a novel quantum theory of an ideal gas,

> based on the hypothesis of a far-reaching formal relation between radiation and gas. According to this theory, the degenerate gas deviates from the gas of [classical] statistical mechanics in a way that is analogous to that in which radiation obeying Planck's law deviates from radiation obeying Wien's law. If Bose's derivation of Planck's radiation formula [given in Bose 1924] is taken seriously, then this theory of the ideal gas is inevitable; for if it is justified to regard radiation as a quantum gas, then the analogy between quantum gas and gas of molecules must be complete." (Einstein 1924a, p. 3)

His third paper on the subject reiterates this point:

> This theory seems justified if one starts with the conviction that (apart from its property of being polarized) a light quantum differs essentially from a monatomic molecule only in that its rest mass is vanishingly small. (Einstein 1925, p. 18)

Einstein had discussed a formal analogy between matter and radiation as early as 1901. In Einstein 1905a, he showed that "radiation obeying Wien's law" behaves thermodynamically as if it were composed of classical particles and proposed his light quantum hypothesis. He had been pondering the significance of Planck's law since 1901, and Einstein 1906 gives a derivation of it showing that "Planck's theory implicitly makes use of the . . . light quantum hypothesis" (p. 199). Why then did he not hit upon

Einstein Studies, vol. 8: Einstein: The Formative Years, pp. 231–251.

the idea of a quantum gas and apply it to both massless light quanta and massive particles well before receiving Bose's work in 1924?[1]

I regard such counterfactual questions as useful: consideration of alternate historical scenarios can teach us a lot about the actual course of history (Stachel 1994a). In the present case, as in *The Hound of the Baskervilles*, the important question may be why the dog did *not* bark. So, in an attempt to shed more light on the precise meaning of his 1905 light quantum hypothesis, I shall offer some speculations on why Einstein did *not* develop the theory of an ideal quantum gas by 1909 .

Those familiar with the work of Klein (see especially 1963, 1964, 1977), Kuhn (1978, chapter 7), and Darrigol (1988, 1991) on Einstein and blackbody radiation will realize my debt to them for many fundamental insights into Einstein's early work on the quantum that are taken for granted here. However, none of them raised this question. Neither did Mehra and Rechenberg but, as we shall see, their discussion suggests a significant new insight into the influence of Wien's work on Einstein that bears directly on this question.[2]

Einstein 1925 mentions two possible objections to his proposed theory of a quantum gas. As applied to blackbody radiation, Planck's formula for the density of blackbody radiation,

$$\varrho(\nu,T) = \frac{8\pi h\nu^3}{c^3[\exp(h\nu/kT)-1]},$$

may be split into two factors, each of which is related to one of the two objections:

(1) The first objection is related to the already-mentioned matter-radiation analogy, which Einstein notes is not universally accepted. (He may well have had Bohr in mind, who refused to countenance the light quantum hypothesis–see Stachel 1999). It is related to the factor $8\pi\nu^2/c^3$, which Bose showed can be interpreted as the number of cells in the phase space of a gas of light quanta, treated as particles.[3]
(2) The second objection is to the method of counting equiprobable cases for a quantum gas, massless or massive:

> The statistical method applied by Bose and myself is by no means indubitable, but on the contrary only appears to be justified *a posteriori* by its success in the case of radiation. (Einstein 1925, p.18)

It is related to the factor $h\nu/[\exp(h\nu/kT)-1]$, which Bose interprets as the average energy per cell when a light quantum gas is in thermal equilibrium. Its value depends on the method of counting equiprobable distributions of light quanta among the cells. From this count one computes the probability W and hence (using Boltzmann's principle) the entropy S of the state of a gas, which is then maximized to determine the thermal equilibrium state.

By examining the role of these two issues in Einstein's work between 1901 and 1909, we may hope to understand why he did not develop the idea of a quantum gas; his ready acceptance and quick application of the idea to massive particles after Bose developed it for light quanta; and why he felt much still remained obscure about the reasons for the idea's success.

2. The Matter-Radiation Analogy

Einstein's speculations on a possible analogy between matter and radiation can be traced back to 1901. Just after he graduated from the Zurich Polytechnic, he wrote his fiancée, Mileva Marić:

> Recently the idea came to me that in the production of light perhaps a direct transformation of motional energy into light takes place due to the parallelism kinetic energy of molecules—absolute temperature—spectrum (spatial radiant energy in the state of equilibrium). (Einstein to Marić, 30 April 1901, Einstein 1987, Doc. 101, p. 295)

Since he was already convinced of the correctness of atomism in the case of matter, such an analogy would suggest the desirability of a similar model of radiation.

He developed such a model in Einstein 1905a, his first light quantum paper, which starts by stressing

> the deep-going formal difference . . . between the theoretical representations physicists have formed of gases and other ponderable matter, and that of Maxwell's theory of electromagnetic processes in so-called empty space.

While the representation of a state of matter only involves specification of

> the positions and velocities of a large but still finite number of atoms and electrons, we require continuous spatial functions for the determination of the electromagnetic state of a space, so that a finite number of quantities is not to

> be regarded as sufficient for determining the electromagnetic state of a space. (Einstein 1905a, p. 151)

As a consequence of this formal distinction, the energy of matter is supposed to be the sum of a finite number of terms, namely a sum over energies associated with the particles of matter, while the energy of the electromagnetic field is supposed to be continuously distributed throughout space. But:

> It seems to me indeed that observations concerning "blackbody radiation," photoluminescence, the generation of cathode rays by ultraviolet light [photoelectric effect] and other groups of phenomena concerned with the generation and transformation of light appear to be better understandable on the assumption that the energy of light is discontinuously distributed in space. According to the assumption contemplated here, the light emitted as radiation from a point is not continuously distributed over greater and greater spaces, but consists of a finite number of energy quanta localized at points of space, which move without dividing and can only be absorbed and emitted as a whole. (p. 151)

Einstein was undoubtedly impelled to make this assumption by the phenomena connected with the emission and absorption of radiation that he mentions. But he was also impelled by his conviction that the ether concept, which underlay the traditional concept of the electromagnetic field, must be abandoned in order to develop a consistent electrodynamics of moving bodies, a conviction he soon published in his first paper on relativity(Einstein 1905b). As he put it in 1910: "Without the ether, energy continuously distributed throughout space seems to me an absurdity (*Unding*)" (Einstein 1910a, p. 177). He raised this point as a strong argument in favor of his conviction that the quantum hypothesis could not be restricted to the material entities interacting with the radiation field, as Planck and others still hoped, but must apply to the field itself, as suggested in Einstein 1905a. The main argument "concerning 'blackbody radiation'" in this paper consists of a demonstration that:

> Monochromatic radiation of low density (within the range of validity of Wien's radiation formula [i.e., when ν/T is large and hence, in modern notation, $\rho(\nu,T)= (8\pi h\nu/c^3)\exp(-h\nu/kT)$], behaves thermodynamically as if it consisted of quanta of energy that are independent of each other and of the magnitude [$h\nu$ in modern notation]. (Einstein 1905a, p. 161)[4]

Einstein shows this by an ingenious inversion of Boltzmann's principle (Einstein gave $S = k \ln W$ this name), a trick he often used thereafter. Instead of reading off the entropy S from a calculation of the probability W, he read off the probability W from the known expression for the entropy S of radiation obeying Wien's law. This leads to an expression for the probability W that, at any moment, the total energy E of the radiation is concentrated in a subvolume υ of the total volume V:

$$W = (\upsilon/V)^{E/h\nu}.$$

Now he introduces the analogy between radiation and matter, by considering a gas consisting of n particles in thermal equilibrium in a similar volume V. If each particle moves according to any law of motion that does not pick out a preferred point or direction in V, then the probability w that, at a given moment, that particle will be in the subvolume υ is clearly: $w = (\upsilon/V)$. If the particles are statistically independent of each other (as are classical particles in an ideal gas), then the probability that, at some moment, all n particles are in υ is clearly the product of the probabilities for each of the n particles:

$$W = w^n = (\upsilon/V)^n.$$

If one takes the analogy between radiation and matter seriously, then comparison of the two formulae for W yields Einstein's conclusion: Radiation in the Wien region behaves like a gas of independent particles, each with energy $h\nu$.[5]

Einstein's argument is easily inverted (and indeed was often interpreted by his contemporaries in this sense): a gas of independent light quanta in thermal equilibrium (its state of maximum entropy) obeys Wien's radiation law.[6] Then one might wonder why didn't Einstein attempt to develop a similar analogy based on Planck's radiation formula?[7] Why did he not look for a way of modifying the statistical treatment of a gas of light quanta that leads to the Planck formula when one maximizes the entropy—in other words, why did Einstein himself not develop Bose's statistics? Neither Kuhn nor Darrigol tries to explain why Einstein initially confined himself to Wien's law, and Klein just hazards a guess: "Einstein based his calculations on this Wien distribution, perhaps because of its greater simplicity" (Klein 1977, p. 24, repeating Klein 1963, p. 68).

3. Why Wien?

Mehra and Rechenberg 1982 offers a detailed discussion of this question, drawing attention to Wien 1900, a previously-neglected source of Einstein's approach in the 1905 paper; but I do not entirely agree with their interpretation of Wien's paper. It was written in response to the measurements of Lummer and Pringsheim, which made it clear that Wien's radiation formula fails at short wave lengths. In his response, Wien stresses that he always expected a difference between the behavior of blackbody radiation at short and long wavelengths.[8] Mehra and Rechenberg claim that Wien expected the departure from classical results at short wavelengths to be due to the failure of classical electrodynamics at these wavelengths.[9] But, on my reading, Wien attributed this departure to the atomistic structure of matter. Since Wien 1900 is not well known, and Mehra and Rechenberg do not cite it at any length, I shall do so:

> I must first of all emphasize that, in contrast to Mr. Planck, I still hold to my earlier-expressed viewpoint, that long and short wavelength electromagnetic waves do not represent just a quantitative difference in their relation to thermal radiation. In the case of absorption, it is indeed generally accepted that, for long wave-lengths, it can be represented by a single vector, or what amounts to the same thing, by assuming the continuity of matter; but in the case of short wave-lengths, on the contrary, the molecular constitution of matter is important. The very same thing must also hold for emission. Therefore, I consider it from the outset as improbable that a radiation law based on molecular assumptions should also be valid for very long wavelengths. The agreement with experiment for short wave-lengths obviously speaks in favor of the approximate validity of the assumptions made [in the derivation of Wien's law] for not-too-long wavelengths.
>
> Bearing this in mind, I consider it not very promising to base a generally-valid radiation formula on molecular hypotheses, as long as a purely thermodynamic derivation is impossible. Rather, the viewpoint of Mr. Paschen, communicated by letter,[10] seems to me more promising, to investigate the representation of the radiation formula for long wavelengths independently of that valid for short wavelengths, just as one uses different formulas for slow electric oscillations than for very rapid ones. (Wien 1900, pp. 537–538)

Mehra and Rechenberg conclude:

> Einstein, in his analysis of the foundations of Planck's radiation formula in 1905, picked up Wien's idea of 1900. That is, he assumed that the heat radiation consisted of two parts: the long-wavelength radiation, described by the

known laws of electrodynamics, and the short-wavelength radiation, determined by laws still unknown. He wanted to discover the nature of these unknown laws. (Mehra and Rechenberg 1982, p. 75)

I agree with these comments, except for two points. The first and more important is the claim, already mentioned, that Einstein's idea of using the Wien formula to "discover . . . these unknown laws" of "short-wavelength radiation" just picked up "Wien's idea." As further (in addition to the quotation above), and hopefully conclusive, evidence that "Wien's idea" was to discover the nature of molecular structure rather than that of radiation, I cite his statement: "I have expressly said that I expect an insight into molecular theory from the testing of the [Wien] law by experiment"(Wien 1900, p. 537).

The other objection is that one cannot describe Einstein's 1905 paper as "an analysis of the foundations of Planck's radiation formula." To the contrary, it is noteworthy how extremely reticent Einstein is in that paper about Planck's radiation formula and the theory behind it.[11] He mentions "the Planck formula" only once (it "satisfies all experiments up to now") in the course of a demonstration that Planck's derivation of Avogadro's number (which Einstein calls the elementary quantum) is "to a certain extent independent of Planck's theory of 'blackbody radiation'"—his only mention of that theory! He is at pains to show that Planck's result depends only on the classical limit of Planck's formula (now called the Rayleigh-Jeans formula), which Einstein had already derived in this paper (see below).

Einstein's reticence concerning Planck's work may be explained by a curious remark near the beginning of his second quantum paper: "On the Theory of Light Production and Absorption" (Einstein 1906). After briefly summarizing his light quantum hypothesis in the first paragraph, he goes on: "At the time [of his 1905 paper] it appeared to me as if Planck's theory of radiation in certain respects was counterposed to my work" (p. 350). In other words, in 1905 he believed that Planck's law was *not* compatible with light quanta, and only in 1906 was he able to give an argument showing "that Planck's theory makes implicit use of the above-mentioned light quantum hypothesis" (p. 350).

Einstein 1905 actually contains an implicit critique of Planck.[12] It shows (p. 44), that the accepted foundations of classical statistical mechanics and Maxwell's electrodynamics lead inevitably to a classical distribution law that is valid (as one would expect on the basis of Wien's argument) for long wave lengths, but must fail for short ones since it leads to a divergence of the total radiant energy. Since Planck claimed to base his work on these same foundations, the clear implication is that he should have arrived at the same law of limited validity.

When Einstein uses a radiation law for short wavelengths to argue from it to the light quantum hypothesis, it is (as noted above) Wien's law: "We take this formula as the basis of our calculations, keeping in mind however that our results are only valid within certain [short-wavelength] limits" (p. 157).

Thus, his theoretical analysis indeed realizes Wien's program of analyzing separately the long- and the short-wavelength behavior of the radiation. But, it was Einstein who made the decisive shift: from Wien's concern with the atomic structure of matter interacting with radiation to an argument for the discontinuous structure of the radiation itself.

What might have suggested this shift from a focus on the matter-radiation interaction to the radiation itself? I suggest it may have been the displacement law. Wien himself notes that, as a result of his displacement law,[13] the wavelength at which his radiation law breaks down "will shift to ever smaller wavelengths with rising temperature"(Wien 1900, p. 537). The same will hold true for the wavelength at which the classical (Rayleigh-Jeans) law breaks down. But if the atomic structure of matter were responsible for this breakdown, one would expect the breakdown to occur at some fixed wavelength related to this atomic structure: the atomic diameter, interatomic distance, or some such parameter.

What has all this to do with Einstein's reasoning in 1905? What evidence is there that he was familiar with Wien's work? One answer is that his first published discussion of blackbody radiation, in Einstein 1904, invokes the displacement law. He investigates energy fluctuations in blackbody radiation, assuming that statistical-mechanical concepts are applicable to this radiation. Using a formula he had developed earlier in the paper for energy fluctuations, he shows that the volume, for which such fluctuations are of the same order of magnitude as the energy average, has a linear dimension that varies inversely with T, with roughly the same numerical coefficient as occurs in Wien's displacement law. Now, if its fluctuations are as big as the energy itself, the concept of a continuous energy distribution breaks down; and Einstein's result, when combined with Wien's displacement law, suggests that this breakdown starts to occur at a certain wavelength associated with the radiation. To someone familiar with Wien's 1900 paper, this might well suggest that the failure of the classical (Rayleigh-Jeans) formula at short wavelengths is associated with a discontinuity in the structure of the radiation, rather than in that of the matter with which it is in equilibrium. But Wien's ideas about the non-classical nature of matter-radiation interactions at short wavelengths may still have played a significant role in shaping Einstein's work: The "heuristic point of view" of its title, as well as all the tests of the light quantum hypothesis suggested in Einstein 1905a, are "concerning the production and transformation [i.e., emission and absorption] of light" by interactions with matter.

What is the likelihood that Einstein read Wien 1900? We know that in 1899 he was so impressed by another work by Wien that he wrote to him (see Einstein 1987, Doc. 57, p. 234). Around 1900, he appears to have read the *Annalen der Physik* regularly; and by 1904–1905 he was evidently familiar with both Wien's formula and displacement law. Taken together with the striking similarity between the research program suggested by Wien 1900 and the one Einstein carried out in 1905—as modified in light of his 1904 result—there is good reason to conjecture that Einstein was familiar with Wien 1900.

4. Elucidating Einstein

This is all I have to add to the well-known discussions of the light quantum hypothesis. Returning to the question raised in the introduction, I can suggest two reasons why Einstein did not immediately develop the idea of a "light quantum gas," that is, a particle model of blackbody radiation:

(1) The analogy between matter and radiation does not demand that one adopt a particle model for radiation;
(2) As indicated above, the particle model of a light quantum gas, taken together with classical statistics (the only kind then known), leads inexorably to the Wien formula.

I shall elaborate on each of these reasons. Einstein's 1905 arguments for light quanta offer no warrant for attributing particle-like properties (other than energy) to them. The argument by analogy with ponderable matter implies that electromagnetic radiation should be modeled as a system whose energy is concentrated at a number of points rather than spread throughout space; but that, of course, allows other possibilities besides a particle model (see below).

His first relativity paper (Einstein 1905b) does contain elements that suggest the possibility of a particle model of radiation. It is surely not a casual remark when he states: "It is noteworthy that the energy and frequency of a light complex change according to the same law with a change in the state of motion of an observer" (Einstein 1905b, p. 299).

But at this stage he was extremely circumspect in mixing quantum and relativistic considerations, and does not remark on the obvious significance of their similar behavior under Lorentz transformations for the possibility of identifying his hypothetical light quantum, for which $E = h\nu$, with such a relativistic "light complex." Nor does he remark on the implication that a light quantum might have momentum, as does a "light complex" according to Maxwell's theory.[14]

Similarly, 1905b notes that, according to relativistic kinematics, "the velocity of light *V* cannot be altered by composition with any 'subluminal velocity'"; but Einstein does not note something that must have been equally obvious to him: This result allows for a relativistic emission theory of light. Taken together with classical kinematics, such an emission theory makes the velocity of light different–and even direction-dependent–in different inertial frames of reference; but Einstein's result shows that relativistic kinematics eliminates this problem. Since emission theories of light were traditionally associated with particle models, this meant that a particle model of light is compatible with special-relativistic kinematics. But again, an emission theory does not require a particle model: other types of emission theory are possible as Ritz's contemporary work showed. Einstein was quite aware that, whatever they were, light quanta could *not* be statistically independent of each other as are "normal" gas particles. For (as mentioned earlier) it follows from reversing the line of argument in the 1905 quantum paper that, if the light quanta *were* independent, the usual probabilistic calculation of the entropy would lead to *Wien's* law for a light quantum gas.

In 1909, Einstein explicitly stated his objections to a particle model of light:

> Indeed, I am not at all of the opinion that one should think of light as composed of quanta localized in relatively small spaces and independent of each other. This would indeed be the most convenient explanation of the Wien end of the radiation formula. But the division of a light ray at the surface of a refracting medium by itself completely forbids this conception. A light ray can divide, but a light quantum cannot divide without change of frequency. (Einstein to Lorentz, 23 May 1909, Einstein 1993b, Doc. 163, p. 193)

5. Einstein's Critique of Complexions

Einstein had strongly-held views about physical probabilities: For him, the time-ensemble definition of the probability of the state of a physical system was primary, as he made clear in 1903:

> Consider a physical system, which can be represented by equations [determining the time rate of change of its state variables as a function of their instantaneous values] and having energy E, during the time interval T starting from an arbitrary initial time. Imagine an arbitrary region Γ of the state variables $p_1 ... p_n$ chosen, so that at a definite moment during the time interval T the variables $p_1 ... p_n$ either lie inside this region or outside it; they will hence lie in the chosen region Γ, during a definite part of the interval T, which we shall call τ. Our condition then takes the form: If $p_1 ... p_n$ are the state variables of a physical system, that is of a system that assumes a stationary state, then the quantity τ/T assumes a definite limiting value for $T = \infty$ for every region Γ. (Einstein 1903, pp. 171–172)

Any *combinatorial* argument, based on counting the number of complexions of the elements composing the system that correspond to a given physical state, must be supplemented by a *dynamical* argument proving that these complexions are equally probable before such a count could be used to define the W needed in Boltzmann's principle. Einstein did not name names in 1905, but made the criticism clear:

> In calculations of entropy based on molecular-kinetic theory, the word "probability" is often used with a meaning that does not agree with the definition of probability given in probability theory. In particular, "cases of equal probability" are often hypothetically postulated in cases where the theoretical model being applied suffices for their deduction, rather than hypothetical postulation. In a separate article, I will show that the so-called "statistical probability" is all that is needed for the treatment of thermal processes, and hope thereby to overcome a logical difficulty that still stands in the way of the application of Boltzmann's principle. (Einstein 1905a, p. 140)[15]

Indeed, his next paper on radiation theory, Einstein 1906, showed that, if one treated Planck's oscillators as systems capable of existing only in states whose energies are integral multiples of $h\nu$, where ν is the frequency of the oscillator, and whose energy changes are discontinuous during absorption and emission of radiation, then the statistical interpretation of Planck's probability W could be maintained.

By 1909, Einstein evidently felt able to criticize Planck openly. After summarizing his time-ensemble definition of the probability of a state, he continues:

> Starting from this definition, one can show that the following equation for the entropy S must hold
>
> $$S = R/N \ln W + \text{const},$$
>
> where the constant is the same for all states of equal energy. Neither Boltzmann nor Planck have given a definition of W. Proceeding quite formally, they set W = the number of complexions of the state in question.
>
> If one now requires that these complexions be equally probable, where the probability of a complexion is defined analogously to how we defined the probability of a state [above], then one arrives at exactly the definition of the probability of a state given [above]; except that the logically unnecessary element complexion has been used in the definition.
>
> Although the relation given between S and W only holds when the probability of a complexion is defined in the way indicated or an equivalent one, neither Boltzmann nor Planck has defined the probability of a complexion. . . . In the oscillator theory of radiation Planck was not free to choose his complexions. He could only postulate the pair of equations
>
> $$S = R/N \ln W$$
>
> and
>
> $$W = \text{number of complexions}$$
>
> if he added the condition that the complexions must be so chosen that, in his theoretical model, they are found to be equally probable on the basis of statistical considerations. In this way, he would have been led to the [Rayleigh-] Jeans formula. (Einstein 1909, pp. 544–545)

Leaving aside the question of whether Einstein's inclusion of Boltzmann in his criticism is justified,[16] Planck's original derivation of the blackbody spectrum applied such combinatorial considerations to the energy of an ensemble of material oscillators in thermal equilibrium with the radiation field, and just assumed all complexions to be equiprobable.

A decade later, Debye applied similar considerations to the radiation field itself, considered as an ensemble of its normal modes confined to a cavity. In both cases, the problem was to divide the fixed total energy into quanta of magnitude $h\nu$ and distribute them among the oscillators–the material oscillators (Planck), or the normal modes of the radiation field

(Debye). A complexion here is just the name for one such way of distributing the total energy. The counting of complexions is based on the assumption that, while material oscillators or normal modes of the radiation field are *distinguishable*, quanta of energy of a given frequency, being just aliquot portions of the total energy at that frequency, are *indistinguishable*. If there are seven quanta of energy associated with one oscillator or normal mode and five with another, it makes no sense to ask *which* seven or *which* five quanta of energy are involved.

Debye, like Planck, just *assumed* that the number of such complexions is a measure of the probability of a state; that is, that each complexion is equally probable. And here Einstein's criticism applies. They offered no dynamical argument why such complexions should be equiprobable in *his* sense (time ensemble) of probability. Indeed, as Einstein pointed out in 1906, Planck had to arbitrarily (from the point of view of classical theory) restrict himself to those complexions that correspond to energies that are integral multiples of $h\nu$ for each frequency ν; and then assume the equiprobability of all such arbitrarily-restricted complexions. As Planck had long recognized, his procedure could only be justified *a posteriori* by its success in producing Planck's formula.[17]

An attempt to turn Debye's argument, based on quanta of energy of the wave field, into an argument based on the model of a gas of particles (light quanta), faces two problems:

(1) How to justify the count of the number of possible states of the gas corresponding to a given energy (or frequency) of a gas particle–this is the problem of the first factor, mentioned in the Introduction.
(2) How to justify the counting of complexions corresponding to a given state of the gas as a counting of equal probabilities–this is the problem of the second factor.

If the first problem is solved, the second only becomes more acute. For, in addition to Einstein's *general* objection to the counting of complexions, a *new* objection arises when such a count of complexions in the Planck-Debye way is reinterpreted as a count of complexions for *particles*. While the energy elements that comprise the total energy are indistinguishable, so that it makes no sense to ask *which* energy quanta are in a given state, how can one avoid this question for particles? Classically, one *cannot*: it seemed part of the nature of the particle concept that such a question makes sense. As was clearly understood at the time (see Ehrenfest 1911), to give it up means giving up the statistical independence of non-interacting particles.

No wonder that even Einstein paused before entering such a statistical morass. Rather than an independent-particle model for light quanta, in 1909 he was searching for a model in which a light quantum represents some sort of singularity in a (nonlinear) electromagnetic field, the motion of which would be "guided" by the equation of motion of the field. Remember that by 1905 he realized that the low-frequency (Rayleigh-Jeans) behavior of the radiation field is correctly described by the application of the equipartition theorem to (low-frequency) classical oscillators in equilibrium with the field; so there is no need to modify the classical wave picture of emission and absorption of radiation for these frequencies. In 1906, Laue had shown that, for coherent bundles of radiation, the addition theorem for entropy must fail if Planck's theory of radiation is to be consistent with the second law of thermodynamics. But if the bundles consisted of independent quanta, the addition theorem would have to hold. Laue considered this a decisive argument against the assumption that the radiation field is composed of light quanta, which he tacitly identified with particles;[18] but Einstein must have taken it as another argument (if one were needed) against the picture of the radiation field as composed of *independent* quanta.

I hope it is now clear that Einstein understood the problems facing an attempt to formulate a quantum gas theory long before he received Bose's paper in 1924. I have already discussed that period and its aftermath in Stachel 1994b, so I shall be brief. Bose rushed in where angels feared to tread, solving the first problem in setting up a theory of a light quantum gas (the counting of states in phase space) very neatly. He solved the second problem (the counting of complexions) without realizing that he was doing something new and revolutionary: He just repeated Planck and Debye's counting method, without realizing that he was abandoning classical (Boltzmann) statistics. Einstein, however, was well aware of what Bose had done, and that this method of counting implied a new, non-classical statistics for light quanta. Since it (like Planck's original method) was justified *a posteriori* by its success in producing Planck's formula, Einstein immediately proceeded to apply the new counting method (i.e., the new statistics that we now call Bose-Einstein) to a quantum gas of massive particles (with the small modification needed in this case, because the particle number is conserved for massive particles). He showed that, as the temperature is lowered, such a gas must undergo a phase change, which is now called Bose-Einstein condensation. Perhaps even more important, he showed that the fluctuations of his quantum gas, like those of the black-body gas, consisted of two terms, a wave term and a particle term; this provided additional evidence to support extension of the ideas of wave-particle duality to massive particles just at the time when DeBroglie and Schrödinger were occupying themselves with such problems.

Einstein was also well aware that the new statistics implies a peculiar statistical entanglement between the particles of his quantum gas. He still hoped to interpret the new statistics as evidence of some novel type of physical interaction between the particles (see Einstein 1924c)–but that is another story, which Don Howard has told elsewhere (see Howard 1990 and Stachel 1997).

6. Appendix: Natanson Neglected?

Kastler 1983 claims that Einstein could have developed the theory of a quantum gas on the basis of Natanson's work, which he overlooked at the time (1911):

> Probably at this time (1911) Einstein's mind was fully occupied by the puzzle of gravitation. Only when he had solved this problem (1916) did he turn his attention again to quantum theory. . . . And so it happened that it took another 13 years [after 1911] before Einstein, inspired by Bose's paper, worked out the statistics of indistinguishable particles and transposed it from photons to atoms and molecules. (Kastler 1983, p. 621)

First of all, it is not correct to say Einstein did nothing in quantum theory between 1911 and 1916 (see the *Collected Papers* vols. 3,4, and 6, Einstein 1993a, 1995, 1996). And even if Kastler were correct about the period before 1916, Natanson was in Berlin in 1915–1916 and Einstein was in contact with him. So it is difficult to accept Kastler's explanation.

Let us look, then, at what Natanson actually did. He was concerned with the problem of counting the number of equiprobable ways of distributing "discrete elements or units of energy" over a number of "receptacles of energy," as he calls them "for brevity." He is quite definite in asserting that these "receptacles" are the "ultimate particles of which matter consists . . . capable only of absorbing, containing and emitting amounts of energy which are multiples of these finite and determinate units" (Natanson 1911a, p. 134). He goes on to point out that the result of such counting depends on whether or not one assumes that "we can identify either receptacles or energy-units" (p. 135). His argument is directed at Planck's earlier derivations of the average energy of a system of such "receptacles of energy," which Planck took to be charged oscillators in equilibrium with the radiation field at some definite temperature. And indeed, when Planck 1911 mentions Natanson 1911b, it is to claim that his method of calculating the average energy of a system of *N* such oscillators is now "completely unambivalent and in particular no longer contains the indefiniteness about which L. Natanson has recently spoken with justification" (Planck 1911, p. 277). So neither Planck nor Natanson was concerned with the radiation field itself, or the distribution of particles (of radiation or of any other kind) over cells in phase space, as Kastler claims:

> He [Natanson] defines the different ways for distributing discrete particles over discrete cells and demonstrates the possibility of three modes of distribution, depending on whether the particles, on the one hand, and the cells, on the other, are indistinguishable or distinguishable entities. (Kastler 1983, p. 620)

Kastler has misidentified Natanson's "energy units" as particles, and Natanson's "ultimate particles" or "receptacles of energy" as cells in phase space. Once we eliminate this confusion, there is no reason to question Einstein's lack of public reaction to Natanson's paper. Even assuming he was familiar with it, it did not remove his objection to combinatorial techniques, old or new, as a way of defining probabilities.

> For his foundation Planck assumes a perfectly definite rule for evaluating the chances; and by a distinct appeal to experience, we find *a posteriori* that the proposition is true. But no one apparently has ever attempted to justify its correctness on general principles and so far as I can see it cannot be done. (Natanson 1911a, p. 139)

As we saw in section 5, as early as 1906 Einstein had found a way of justifying Planck's results by a statistical argument applied to oscillators with discrete energies. And there is no reason why Natanson's work should have suggested that Einstein reconsider his conclusion that a particulate model of radiation must lead to Wien's law. As emphasized above, Natanson was concerned with distribution of energy units over material entities, not light quanta over cells in phase space. Ehrenfest, who did consider the latter problem (see Ehrenfest 1911, Ehrenfest and Onnes 1914), had just demonstrated that Planck's combinatorics is incompatible with the concept of statistically-independent particles–the only kind that Einstein had been considering before Bose came along.

Acknowledgement. I thank Olivier Darrigol for a critical reading of the text.

NOTES

[1] This question first occurred to me in the course of writing Stachel 1994b, which discusses Einstein's reaction to Bose's work.

[2] For their discussion of Einstein's early work, see Mehra and Rechenberg 1982, pp. 59–83.

[3] "$8\pi\nu^2/c^3\, V\, d\nu$ can be interpreted as the number of elementary cells of the six-dimensional phase space for the [light] quanta" (Bose 1924, p. 384). Here V is the volume of the cavity.

[4] Einstein 1905a does not use either Planck's constant h or Boltzmann's constant k; but rather two equivalent constants. This is of some significance to an understanding of Einstein's approach; but this has been elucidated in Klein 1963, so I shall not repeat his discussion. Einstein 1910b, which sketches his 1905 argument again (in some respects more carefully–see the next footnote, does use h, but still uses R/N instead of k.

[5] In 1910, he repeats the argument with a more cautious conclusion, not using the word "particle." After showing that "for sufficiently dilute radiation," $W = \left[\frac{V}{V_0}\right]^{\left[\frac{E_0}{h\nu}\right]}$ is "the probability that at a given instant all the energy E_0 is included in a volume V," he concludes: "one can easily show that this expression is not compatible with the superposition principle. The radiation behaves . . . as if its energy were localized in $E_0/h\nu$ points moving independently of each other" (Einstein 1910b, p. 251).

[6] Renn 1993, p. 334, has emphasized the invertibility of Einstein's 1905 argument.

[7] As is well known, when he did discuss energy fluctuations in the Planck distribution in Einstein 1909, he showed that they are the sum of the two terms: one could be interpreted as a particle term and the other as a wave term. While it is not impossible that he first made this calculation much earlier, as is suggested by some comments in his "Autobiographical Notes" (see Einstein 1979, p. 48), in the next section we shall argue that initially he was suspicious of Planck's formula and had good reason to start with Wien's formula. This argument replaces my earlier speculation in Stachel 1994b, that Einstein did attempt such a derivation.

[8] In support of this claim he cites Wien 1893b which argues on the basis of the molecular structure of matter that there should be an upper limit to the wave lengths that occur in the blackbody radiation from rigid bodies. It is possible that Einstein followed up this reference, but this point is not important for my argument.

[9] "Wien had stated . . . that the heat radiation satisfying his law could not be explained on the basis of the electrodynamic theory; rather it exhibited a structure which was different in quality from that of the long-wavelength radiation, the radiation which obeyed the laws of classical electrodynamics" (Mehra and

Rechenberg 1982, p. 75).

[10] It would be interesting to search for this letter, or other evidence of Paschen's views. Lacking this, I shall here refer to Wien's views, leaving open the question of how far they reflect Paschen's.

[11] Klein 1977, p. 29 noted this, as do the Editors of the Einstein Papers: "Among Einstein's papers on the quantum hypothesis, the 1905 paper is unique in arguing for the notion of light quanta without using either the formal apparatus of his statistical papers or Planck's law" (see the headnote, "Einstein's Early Work on the Quantum Hypothesis," Einstein 1989, p. 139).

[12] I agree with Darrigol 1988, which interprets a later remark of Besso as a claim that in 1904–1905 he persuaded Einstein not to criticize Planck explicitly (p. 57). Below we shall quote another passage from Einstein 1905a, cited by Darrigol as an implicit criticism of Planck.

[13] Wien 1894 shows by a thermodynamic argument that the energy spectrum must be of the form $\lambda^{-5}f(\lambda T)$ where λ is the wavelength and T the absolute temperature of the radiation. Thus, any particular wavelength characterizing the spectrum, such as λ_m , the wavelength at the energy maximum, will vary as $1/T$ times a universal coefficient, a result that became known as Wien's displacement law.

[14] It was only in 1909 that Einstein gave the first of a series of arguments that culminated in 1916 with the definite attribution of momentum to light quanta.

[15] I agree with Darrigol 1988, which sees the omission of Planck's name as a strategic move by Einstein. See note 12.

[16] See the paper by Jürgen Renn in this volume for a discussion of Einstein's knowledge of Boltzmann's work.

[17] In 1897, he wrote: "The probability of each [macro-]state is proportional to the number of complexions corresponding to it, or, in other words, any particular complexion has the same probability as any other complexion. . . . This statement is the central point of the whole theory under discussion; it can be proved in the final analysis only by experience" (cited from Kastler 1983, p. 616).

[18] See Max Laue to Albert Einstein, 2 February 1906, Einstein 1993, Doc. 37, pp. 41–42.

REFERENCES

Bose, Satyendra Nath (1924). "Wärmegleichgewicht im Strahlungsfeld bei Anwesenheit der Materie." *Zeitschrift für Physik* 27: 384–392.

Darrigol, Oliver (1988). "Statistics and Combinatorics in Early Quantum Theory." *Historical Studies in the Physical and Biological Sciences* 19: 1–80.

— (1991). "Statistics and Combinatorics in Early Quantum Theory, II: Early Symptoms of Indistinguishability and Holism." *Historical Studies in the Physical and Biological Sciences* 21: 237–298.

Debye, Peter (1910). "Der Wahrscheinlichkeitsbegriff in der Theorie der Strahlung." *Annalen der Physik* 33: 1427–1434.

Ehrenfest, Paul (1911). "Welche Züge der Lichtquantenhypothese spielen in der Theorie der Wärmestrahlung eine wesentliche Rolle." *Annalen der Physik* 36: 91–118.

Ehrenfest, Paul and Kamerlingh Onnes, Heike (1914). "Vereinfachte Ableitung der kombinatorischen Formel, welche der Planckschen Strahlungstheorie zugrunde liegt." *Annalen der Physik* 46:1021–1024.

Einstein, Albert (1903). "Eine Theorie der Grundlagen der Thermodynamik." *Annalen der Physik* 11: 170–187. Reprinted in Einstein 1989, Doc. 4, pp. 77–94.

— (1904). "Zur allgemeinen molekularen Theorie der Wärme." *Annalen der Physik* 14: 354–362. Reprinted in Einstein 1989, Doc. 4, pp. 99–107.

— (1905a). "Über einen die Erzegung und Verwandlung des Lichtes betreffenden heuristischen Gesichtspunkt." *Annalen der Physik* 17: 132–148. Reprinted in Einstein 1989, Doc. 14, pp. 150–166.

— (1905b). "Zur Elektrodynamik bewegter Körper." *Annalen der Physik* 17: 891–921. Reprinted in Einstein 1989, Doc. 23, pp. 276–306.

— (1906). "Zur Theorie der Lichterzeugung und Lichtabsorption." *Annalen der Physik* 20: 199–206. Reprinted in Einstein 1989, Doc. 34, pp. 350–357.

— (1909). "Zum gegenwärtigen Stand des Strahlungsproblems." *Physikalische Zeitschrift* 10: 185–193. Reprinted in Einstein 1989, Doc. 56, pp. 542–550.

— (1910a). "Response to Manuscript of Planck." Written before 18 January 1910. In Einstein 1993a, Doc. 3, pp. 177–178.

— (1910b). "Sur la théorie des quantités lumineuses et la question de la localisation de l'énergie électromagnétique." *Archives des sciences physiques et naturelles* 29: 525–528. Reprinted in Einstein 1993a, Doc. 5, pp. 249–252.

— (1924a). "Quantentheorie des einatomigen idealen Gases." *Preussische Akademie der Wissenschaften* (Berlin). *Physikalisch-mathematische Klasse. Sitzungsberichte*: 261–267.

— (1924b). "Quantentheorie des einatomigen idealen Gases. Zweite Abhandlung." *Preussische Akademie der Wissenschaften* (Berlin). *Physikalisch-mathematische Klasse. Sitzungsberichte* (1925): 3–14.

— (1924c). "Über den Aether." *Schweizerische naturforschende Gesellschaft. Verhandlungen* 105: 85–93.

— (1925). "Quantentheorie des idealen Gases." *Preussische Akademie der Wissenschaften* (Berlin). *Physikalisch-mathematische Klasse. Sitzungsberichte*: 18–25.

— (1979). *Autobiographical Notes*. Paul Arthur Schilpp, transl. and ed. (La Salle, IL and Chicago: Open Court).

— (1987). *The Collected Papers of Albert Einstein.* Vol. 1, *The Early Years, 1879–1902*. John Stachel, et al., eds. Princeton: Princeton University Press.

— (1989). *The Collected Papers of Albert Einstein.* Vol. 2, *The Swiss Years: Writings, 1900–1909.* John Stachel, et al., eds. Princeton: Princeton University Press.

— (1993a). *The Collected Papers of Albert Einstein.* Vol. 3, *The Swiss Years: Writings, 1909–1911.* Martin J. Klein, et al., eds. Princeton: Princeton University Press.

— (1993b). *The Collected Papers of Albert Einstein.* Vol. 5, *The Swiss Years: Correspondence, 1902–1914.* Martin J. Klein, et al., eds. Princeton: Princeton University Press.

— (1995). *The Collected Papers of Albert Einstein*, vol. 4, *The Swiss Years: Writings, 1912–1914.* Martin J. Klein, et al., eds. Princeton: Princeton University Press.

— (1996). *The Collected Papers of Albert Einstein*, vol. 6, *The Berlin Years: Writings 1914–1917.* Martin J. Klein, et al., eds. Princeton: Princeton University Press.

Howard, Don (1990). "'Nicht sein kann was nicht sein darf,' or the Prehistory of EPR, 1909–1935: Einstein's Early Worries About the Quantum Mechanics of Composite Systems." In *Sixty-two Years of Uncertainty: Historical, Philosophical, and Physical Inquiries into the Foundations of Quantum Physics.* Arthur Miller, ed. New York: Plenum, pp. 61–111.

Kastler, Alfred (1983). "On the Historical Development of the Indistinguishability Concept for Microparticles." In *Old and New Questions in Physics, Cosmology, Philosophy, and Theoretical Biology.* Alwyn van der Merwe, ed. New York and London: Plenum, pp. 607–623.

Klein, Martin (1963). "Einstein's First Paper on Quanta." *The Natural Philosopher.* Vol. 1. Daniel E. Gershenson and Daniel A. Greenberg, eds. New York: Blaisdel, pp. 59–85.

— (1964). "Einstein and the Wave-Particle Duality."*The Natural Philosopher.* Vol. 3. Daniel E. Gershenson and Daniel A. Greenberg, eds. New York: Blaisdell, pp. 3–49.

— (1970). *Paul Ehrenfest.* Vol. 1, *The Making of a Theoretical Physicist* Amsterdam: North-Holland.

— (1977). "The Beginnings of the Quantum Theory." In *Storia della Fisica del XX Secolo. Rendiconti della Scuola Internazionale di Fisica "Enrico Fermi."* LVII Corso. Charles Wiener, ed. New York and London: Academic, pp. 1–39.

Kuhn, Thomas S. (1978). *Black-Body Theory and The Quantum Discontinuity, 1894–1912.* Oxford: Clarendon Press; New York: Oxford University Press.

Laue, Max (1906). "Zur Thermodynamik der Interferenzerscheinungen." *Annalen der Physik* 20: 365–378.

Mehra, Jagdish and Rechenberg, Helmut (1982). *The Historical Development of Quantum Theory.* Vol 1, part 1, *The Quantum Theory of Planck, Einstein, Bohr and Sommerfeld: Its Foundation and the Rise of its Difficulties 1900–1925.* New York, Heidelberg, and Berlin: Springer-Verlag.

Natanson, Ladisłas (1911a). "On the Statistical Theory of Radiation." *Bulletin de l'Academie des Sciences de Cracovie (A)*: 134–148.

— (1911b). "Über die statistische Theorie der Strahlung." *Physikalische Zeitschrift* 12: 659–665 (German translation of Natanson 1911a).

Pais, Abraham (1982). *'Subtle is the Lord...' The Science and the Life of Albert Einstein.* Oxford: Clarendon Press; New York: Oxford University Press.

Planck, Max (1900). "Zur Theorie des Gesetzes der Energieverteilung im Normalspectrum." *Deutsche physikalische Gesellschaft* Berlin. *Verhandlungen* 2: 237–245.

— (1911). "Die Gesetze der Wärmestrahlung und die Hypothese der elementaren Wirkungsquanten." Report to the 1911 Solvay Congress. Cited from: Max Planck. *Physikalische Abhandlungen und Vorträge*, vol. 2. Braunschweig: Friedrich Vieweg & Sohn, 1958, pp. 269–286.

Renn, Jürgen (1993). "Einstein as a Disciple of Galileo: A Comparative Study of Concept Development in Physics." *Science in Context* 6: 311–341.

Stachel, John (1986). "Einstein and the Quantum: Fifty Years of Struggle." In *From Quarks to Quasars: Philosophical Problems of Modern Physics*. Robert Colodny, ed. Pittsburgh: University of Pittsburgh Press, pp. 349–385.

— (1994a). "Scientific Discoveries as Historical Artifacts." In *Trends in the Historiography of Science*. Kostas Gavroglu, ed. Dordrecht: Kluwer, pp. 139–148.

— (1994b). "Einstein and Bose." In *Bose and 20th Century Physics*. Partha Ghose, ed. Dordrecht: Kluwer (forthcoming).

— (1997). "Feynman Paths and Quantum Entanglement: Is There Any More to the Mystery?" In *Potentiality, Entanglement, and Passion-at-a-Distance.* R. S. Cohen et al., eds. Dordrecht: Kluwer, pp. 245–255.

— (1999). "Bohr and the Photon." In *Going Critical: Selected Papers*. Dordrecht: Kluwer (forthcoming).

Wien, Wilhelm (1893a). "Eine neue Beziehung der Strahlung schwarzer Körper zum zweiten Hauptsatz der Wärmetheorie." *Königlich Preussische Akademie der Wissenschaften* (Berlin). *Sitzungsberichte*: 55–62.

— (1893b). "Die obere Grenze der Wellenlängen, welche in der Wärmestrahlung fester Körper vorkommen können; Folgerungen aus dem zweiten Hauptsatz der Wärmetheorie." *Annalen der Physik* 49: 633–641.

— (1894). "Temperatur und Entropie der Strahlung." *Annalen der Physik* 52: 132–165.

— (1896). "Über die Energievertheilung im Emissionsspectrum eines schwarzen Körpers." *Annalen der Physik* 58: 662–669.

— (1900). "Zur Theorie der Strahlung schwarzer Körper. Kritisches." *Annalen der Physik* 3: 530–539.

Index

Abbe, Ernst, 43, 46, 47, 78
Abraham, Max, 176, 193
Ampère, André-Marie, 6
Aristotle, 34
Auerbach, Felix, 46
Avenarius, Richard, 6
Avogadro's number, 125, 126, 147, 149, 183, 211, 218, 235

Bachelier, Louis, 209, 215, 217
Beller, Mara, 1, 14, 83
Bennett, G.J., 216
Bernoulli, Daniel, 62
Bernstein, Aaron (a.k.a. A. Rebenstein), 14, 23-40, 70
Besso, Michele, 5, 6, 7, 14, 15, 17, 66, 130, 161, 192, 194
Bierhalter, Günter, 151, 152
Blackmore, John T., 213
Bohr, Niels, 232
Boltzmann, Ludwig, 6, 10, 12, 13, 16, 44, 67, 68, 69, 97, 107, 108, 109, 111, 112, 117-121, 123,124, 126, 127, 128, 132–146, 148–152, 161, 176, 180, 194, 195, 207, 208, 233, 235, 241, 242, 244, 247, 248
Born, Max, 85
Bose, Satyendra Nath, 231, 232, 233, 235, 244, 247
Bradley, James, 32
Brush, Steven, 149
Buchdahl, Gerd, 90, 92, 100, 101
Bucherer, Abraham, 91
Büchner, Ludwig, 14, 23, 34, 39
Buchwald, Jed Z., 148

Cahan, David, 46, 51, 52, 55, 58
Carnap, Rudolf, 221, 226
Cartwright, Nancy, 223
Cassirer, Ernst, 87
Chandrasekhar, Subrahmanyan, 216, 219, 220, 225
Clausius, Rudolf, 150, 151
Clifford, William Kingdon, 6, 17
Coffa, J. Alberto, 102
Comstock, D.F., 195
Crew, Henry, 54, 57, 61, 77

Damerow, Peter, 148, 152
Darrigol, Olivier, 232, 235, 246, 248
Darwin, Charles, 38
DeBroglie, Louis, 245
Debye, Peter, 242, 243
Dedekind, Richard, 6
Deltete, Robert John, 15
Descartes, René, 84, 86, 104
Dirac, Paul, 207
Draper, John William, 51
Drude, Paul, 44, 67, 107, 112, 113, 114, 115, 117–124, 127–132, 134, 145, 150, 152, 166, 195
Ducrue, Joseph, 78

Duhem, Pierre, 83, 92, 94
Dukas, Helen, 1, 16

Earman, John, 194
Ehrat, Jakob, 73, 166
Ehrenfest, Paul, 107, 109, 110, 112, 140, 141, 150, 182, 184, 195, 197, 246
Ehrenfest, Tatiana, 107, 109, 110, 140, 141, 150
Einstein, Albert,
  and the aether, 3, 9, 17, 165, 166, 167, 172, 175, 177, 178, 179, 181, 187, 188, 193
  and atomism, 1,3, 10, 24, 27, 40, 42, 44, 98, 103, 107, 112, 113, 124–127, 129, 130, 132, 143–147, 207, 213, 233, 236, 237
  and *Autobiographical Notes*, 23, 83, 177, 179, 187, 188, 192, 195
  and Bose-Einstein statistics, 244, 245, 246
  and Brownian motion, 1, 7, 9, 12, 109, 146, 161, 203, 204, 205, 207, 213–217, 219, 221, 222, 223
  and causality, 98, 99, 100, 101, 104
  and Chapman-Kolmogorov equation, 215
  and the concept of God, 91, 91, 91, 100, 101
  and conventionalism, 92, 94, 100
  and determinism, 99, 101, 104
  and doctoral dissertation, 9, 116, 117, 120, 121, 122, 123, 160, 161, 210
  and the electron, 107, 111, 112, 114, 115, 121, 122, 123, 127, 128, 130, 131, 132, 134, 136, 145, 149, 150, 152, 186, 187, 188, 189, 191, 192, 195
  and empiricism, 86, 87, 89, 90, 100, 103
  and epistemology, 6, 83, 86, 87, 91, 92, 93
  and experiment, 3, 9, 15, 64, 67, 73, 76, 77, 79, 84, 87, 90, 91, 98, 103, 104, 113, 116, 127, 128, 129, 131, 146, 161, 162, 165, 166, 167, 176–181, 183, 184, 185, 192, 193, 207, 213, 225, 237
  and emission theory of light, 127, 168, 178, 179, 182, 183, 184, 189, 193, 196, 240
  and light quanta, 13, 14, 123, 147, 148, 176–187, 189, 196, 231, 232, 233, 234, 237–240, 243, 244, 246, 247, 248
  and the light postulate, 163, 167, 168, 174, 183, 184, 185, 191
  and logical positivism, 86, 91, 98, 101, 102, 104
  and nature of theories, 10, 11,12, 13, 14, 17, 18, 91, 92, 93, 94, 95, 96
  and neo-Kantianism, 84, 87, 89, 101, 102
  and "On the Electrodynamics of Moving Bodies," 160, 161, 162, 170, 171, 172, 174, 192, 193
  and popularization of science, 23, 24
  and positivism, 83, 86, 98, 101, 104
  and principle of relativity, 9, 13,14, 110, 163, 164, 166–171, 173, 174, 177–181, 185, 186, 191–194
  and rationalism, 83, 84, 86, 87, 88, 89, 91
  and realism, 92, 98, 99, 100, 104
  and relativity of simultaneity, 160, 161, 162, 168, 173, 174, 187, 192
  and religious development, 38
  and Stoke's formula, 214
  and theory of radiation, 4, 8, 11, 12, 13, 107, 108, 111, 112, 114, 127, 133, 134, 147, 148, 174–181, 183, 231–244
  and time concept, 7, 11, 98, 111, 147, 162, 163, 168, 169, 173, 192
  and underdetermination, 94
Einstein, Hermann (father), 74, 76
Einstein, Jakob (uncle), 38, 76
Euler, Leonhard, 62
Exner, Franz, 207, 210

Faraday, Michael, 170, 177, 178
Fehr, Johannes, 148

Fessenden, Reginald Aubrey, 225
Fine, Arthur, 98, 99, 100, 101
Fisch, Adolf, 66
Fischer, Kuno, 78
FitzGerald, George Frances, 94, 191
Fizeau, Armand, 31, 32, 165, 192, 193
Foerster, Wilhelm, 58, 59
Fokker, Adriann Daniel, 215, 217, 225
Föppl, August, 44, 67, 164, 165
Forman, Paul, 57
Foucault, Léon, 32
Fraassen, Bas van, 100, 103, 226
Frank, Phillip, 165, 195
Freud, Sigmund, 102
Frey-Wyßling, A., 53
Friedman, Michael, 91, 102
Fürth, Reinhold, 214, 215, 225

Galilei, Galileo, 4, 7, 10, 89, 169, 191, 192
Gallavotti, Giovanni, 148, 149, 150, 152
Gardiner, Crispin W., 221, 225
Gauss, Carl Friedrich, 28
Geiser, Carl Friedrich, 54, 55
Gibbs, Josiah Willard, 71, 109, 112, 133, 140–144, 152
Giedymin, Jerzy, 17
Glymour, Clark, 190, 194
Gmür, Annette, 102
Gmür, Nick, 102
Goethe, Johann Wolfgang von, 35, 40, 102
Goldberg, Stanley, 102
Gregory, Frederick, 1, 14, 23
Grossman, Marcel, 2, 73, 75, 76, 77, 117, 119, 120, 126, 149, 161, 166
Gruner, Paul, 130
Gugerli, David, 148
Guggenbühl, Gottfried, 54, 55, 56, 63
Gumlich, Ernst, 69

Habicht, Conrad (Konrad), 4, 6, 16
Harman, P.M., 147, 149
Hasenöhrl, Fritz, 176
Häusermann, Elsi, 53
Hegel, Georg W. F., 34, 39

Heilbron, John, 57
Helmholtz, Anna von (2nd wife), 58, 59, 61
Helmholtz, Hermann von, 6, 16, 43–51, 55, 56, 58–63, 67, 69, 70, 71, 74, 77, 83, 85, 138, 146, 151, 152
Helmholtz, Robert (son), 58
Hentschel, Klaus, 68, 77
Hermann, Armin, 132, 194
Herneck, Friedrich, 1, 5, 16
Hertz, Heinrich, 2, 6, 16, 44, 53, 61, 66, 67, 69, 70, 71, 78, 171, 172, 175, 189, 191
Hertz, Paul, 152
Herzog, Albin, 67, 75
Hilbert, David, 195
Hines, C.O., 196, 197
Hoffman, Banesh, 1
Höflechner, Werner, 150
Holborn, Ludwig, 69
Holton, Gerald, 14, 15, 16 , 83, 88, 91, 92, 100, 102, 103, 169, 170, 171, 193, 195
Howard, Don, 77, 84, 87, 92, 94, 99, 100, 102, 103, 193, 222, 224, 245
Humboldt, Alexander von, 14, 23, 24, 40
Hume, David, 6, 7, 16, 17, 76, 83, 85, 89, 96, 103
Hurwitz, Adolf, 43, 63, 74

Infeld, Leopold, 169, 195
Ishiwara, Jun, 162, 166
Itô, Kiyoshi, 220

Jacobi, Carl Gustav, 140
Jaeger, Wilhelm, 69
Jesus Christ, 102
Jungnickel, Christa, 122

Kac, Mark, 217
Kangro, Hans, 51
Kant, Immanuel, 14, 76, 83–95, 98–104
Kappeler, Karl, 48, 49, 54, 55, 57
Kaiser, Walter, 128, 132
Kamerlingh Onnes, Heike, 130, 246
Kastler, Alfred, 245, 246, 248

Kaufmann, Walter, 91, 128
Kayser, Heinrich, 61
Kayser, Rudolph (a.k.a. Anton Reiser), 121, 162, 165, 166, 195
Keesom, Willem Hendrik, 130
Kipling, Rudyard, 225
Kirchhoff, Gustav, 6, 16, 44, 46, 47, 67, 69, 210
Kitcher, Philip, 93, 104
Klein, Martin, 8, 14, 71, 92, 96, 97, 112, 151, 153, 193, 232, 235, 247, 248
Kleiner, Alfred, 6, 10, 16, 17, 117, 120–123, 149, 161, 166
Kluyver, J.C., 216, 217
Kohlrausch, Friedrich, 49
Köhnke, Klaus C., 102
Kollros, Louis, 66, 67, 73, 76, 77
Körber, Hans-Günther, 74
Kragh, Helge, 74
Kreichgauer, Damian, 69
Kremer, Richard L., 51
Kuhn, Thomas S., 232, 235, 248
Kunz, J., 195

Langer, William, 23
Langevin, Paul, 205, 213, 215–221, 225
Lasius, G., 56
Laub, Jakob, 10
Laue, Max von, 177, 244, 248
Leggett, A.J., 203, 205, 206, 222, 223, 224
Leverrier, Urbain, 26
Levins, Richard, 223
Lindeck, Stephan, 69
Lindemann, Ferdinand, 17
Linnaeus, Carolus, 93
Liouville, Joseph, 145
Lorentz, Hendrik Antoon, 3, 4, 7, 9, 14, 91, 96, 110, 111, 124, 129, 130, 163, 164, 166, 167, 169, 170, 173, 174, 175, 186, 187, 189–192, 194, 195, 197, 240
Lorenz, Richard, 217
Lübsen, Heinrich Borchert, 38
Lummer, Otto, 236

MacLane, Saunders, 226
McCormmach, Russell, 57, 72, 123
Mach, Ernst, 3, 5, 6, 7, 15, 16, 17, 44, 69, 76, 83, 86, 96, 98, 104, 207, 213
Marez Oyens, Felix de, 149
Marić, Mileva, 5, 6, 15, 16, 17, 64, 68, 69, 70, 74, 75, 77, 113–123, 127–129, 132, 145, 149, 150, 159, 161, 165, 166, 233
Markov, Andrei Andreevich, 211, 215
Maupertius, Pierre-Louis Moreau de, 93
Maxwell, James Clerk, 2–5, 7, 9–13, 15, 44, 53, 66, 67, 73, 92, 93, 107, 108, 109, 119, 124, 140, 142, 146, 149, 152, 163, 164, 170, 171, 172, 175, 182, 183, 186, 187, 189, 190, 191, 194, 195, 208, 233, 237, 240
Mayer, Robert, 35
Mehra, Jagdish, 232, 236, 237, 247, 248
Meyer, Oskar Emil, 58, 68
Michelson, Albert, 51, 98, 104, 110, 165, 167
Mill, John Stuart, 6, 16
Miller, Arthur, 91, 102
Minkowski, Hermann, 10, 43, 67
Morley, Edward W., 98, 104, 110, 165, 167
Mousson, Albert, 57
Moyal, J.E., 215
Mühsam, Hans, 24, 30

Nabl, Josef, 123, 145, 146
Natanson, Ladislas, 245, 246
Neumann, Franz, 58
Newton, Isaac, 2, 4, 7, 8, 9, 28, 33, 93, 151, 164
Nicolas, Marcel, 25
Nicolsen, Malcolm, 40
Norton, John, 193
Nye, Mary Jo, 15, 147, 213

Oechsli, Wilhelm, 53, 54, 55, 63
Olesko, Kathryn M., 58, 77
Ornstein, Leonard, 214, 220
Ostwald, Wilhelm, 74, 213

Pais, Abraham, 72
Paschen, Friedrich, 236, 248
Pauli, Wolfgang, 195
Pearson, Karl, 6, 204, 216, 217
Pernod, Johannes, 44, 57–66, 68, 69, 73, 77
Perrin, Jean, 146, 147, 213, 217, 225
Plank, Max, 4, 5, 11, 12, 13, 15, 51, 103, 115, 127, 145, 177, 179, 184, 187, 195, 196, 225, 231, 232, 234–238, 241–248
Platter, Julius, 85, 103
Poggendorf, J.C., 78
Poincaré, Henri, 6, 7, 10, 16, 17, 83, 85, 103, 194, 209
Pringsheim, Ernst, 236
Pyenson, Lewis, 39, 68, 76

Quine, Willard van Orman, 225, 226

Rayleigh, Lord (a.k.a. John W. Strutt), 49, 208, 209, 216
Rebenstein, A. (a.k.a. Aaron Bernstein), 14, 23-40, 70
Rechenberg, Helmut, 232, 236, 237, 247, 248
Reinganum, Maximilian, 117, 118, 128, 129, 130, 150
Reiser, Anton (a.k.a. Rudolph Kayser), 121, 162, 165, 166, 195
Renn, Jürgen, 7, 107, 113, 122, 149, 150, 193, 247, 248
Riecke, Eduard, 74, 75
Riemann, Bernhard, 6, 16
Rippenbein, A.P., 168
Ritter, Karl, 61
Ritz, Walter, 168, 179, 189, 195, 240
Römer, Olaf, 31
Rosen, Nathan, 16
Ryckman, T.A., 91, 102
Rynasiewicz, Robert, 7, 159, 197

Sarkar, Sahotra, 1, 7, 203, 222, 226
Sauer, Tilman, 148
Savić, Helene, 116, 120
Schaffer, Simon, 39
Schlick, Moritz, 17, 101
Schoepflin, Urs, 148
Schneebeli, Heinrich, 57, 58
Schrödinger, Erwin, 245
Schulmann, Robert, 113, 122, 193
Schweber, Silvan, 148
Seelig, Carl, 16, 66, 69, 73, 116, 159
Seeliger, Rudolf, 131
Shankland, Robert, 168, 171, 195, 196
Shimony, Abner, 222, 224
Siemens, Werner von, 49, 55
Smoluchowski, Marian, 207, 213, 217–220, 223, 225
Solovine, Maurice, 6, 16, 17, 195
Sommerfeld, Arnold, 12, 18, 123, 131, 132, 225
Spencer, Herbert, 87
Speziali, Pierre, 6, 7, 17, 66, 192, 194
Spieker, Theodor, 39
Spinoza, Baruch (Benedict), 6, 16, 83, 84, 86, 88, 89, 94
Stachel, John, 1, 8, 39, 77, 102, 167, 193, 222, 232, 244, 247
Stadler, August, 83–87, 95, 101, 102
Staedel-Schneider, Gudrun, 149
Stark, Johannes, 160, 194
Stein, Howard, 225
Stern, Alfred, 69, 70, 74
Stern, Moritz, 69
Stern, Sigismund, 26
Stewart, O.M., 195
Stokes, George Gabriel, 205, 210, 214
Strutt, John W. Lord Rayleigh, 49, 208, 209, 216
Study, Eduard, 99

Talmey, Max, 23, 24, 38, 39, 84
Teucher, Emil, 71
Thiesen, Max, 58, 61, 62
Thompson, Colin J., 206
Thomson, J.J., 71
Thomson, Silvanus, 49
Thomson, William Lord Kelvin, 49, 52
Tisza, Laszlo, 214
Tolman, Richard, 195
Tuchschmid, August, 73

Uhlenbeck, George E., 220

Violle, Jules, 10, 17, 69
Viscardini, Mario, 168, 182

Wander de Haas-Lorentz, Johannes, 214
Weart, Spencer, 57
Weber, Heinrich Friedrich, 43–69, 71–78, 116, 117, 120, 122, 195
Weber, Wilhelm, 28
Weiss, Pierre, 45, 46, 48, 49, 52, 55, 63, 64, 76, 78
Weltsch, Robert, 25
Wertheimer, Max, 162, 169, 170
Wheaton, Bruce R., 153
Wheatstone, Charles, 31, 49, 72
Wiebe, Hermann, 58, 61, 62
Wiedemann, Gustav, 46, 47
Wien, Wilhelm, 4, 11, 51, 72, 96, 97, 110, 113, 148, 165, 166, 176, 177, 180, 181, 182, 184, 191, 231, 232, 234–240, 246, 247, 248
Wiener, Norbert, 208, 209, 217, 225
Wiener, Otto, 74
Wiernik, Peter, 25, 26, 30
Willey, Thomas E., 102
Wimsatt, William, 222, 226
Winteler, Jost, 118, 122
Winteler-Einstein, Maja, 23

Zangger, Heinrich, 76
Zeisler, E., 98

GPSR Compliance

The European Union's (EU) General Product Safety Regulation (GPSR) is a set of rules that requires consumer products to be safe and our obligations to ensure this.

If you have any concerns about our products, you can contact us on ProductSafety@springernature.com

In case Publisher is established outside the EU, the EU authorized representative is:

Springer Nature Customer Service Center GmbH
Europaplatz 3
69115 Heidelberg, Germany

Batch number: 10725973

Printed by Printforce, the Netherlands